# Advances in Anatomy
# Embryology and Cell Biology

## Vol. 81

Editors
F. Beck, Leicester  W. Hild, Galveston
J. van Limborgh, Amsterdam  R. Ortmann, Köln
J.E. Pauly, Little Rock  T.H. Schiebler, Würzburg

Ursula-Friederike Habenicht
Friedmund Neumann

# Hormonal Regulation of Testicular Descent

With 39 Figures

Springer-Verlag
Berlin Heidelberg New York Tokyo 1983

Dr. Ursula-Friederike Habenicht
Prof. Dr. Friedmund Neumann
Research Laboratories of Schering AG
Department of Endocrine — Pharmacology
Müllerstr. 170—178
D-1000 Berlin 65
FRG

ISBN-13:978-3-642-69123-2     e-ISBN-13:978-3-642-69121-8
DOI: 10.1007/978-3-642-69121-8

Library of Congress Cataloging in Publication Data
Habenicht, Ursula-Friederike, 1952 — Hormonal regulation of testicular
descent. (Advances in anatomy, embryology, and cell biology; v. 81)
Bibliography: p. Includes index.
1. Testis — Addresses, essays, lectures. 2. Hormones, Sex — Physiological
effect — Addresses, essays, lectures. 3. Cryptorchism — Animal models —
Addresses, essays, lectures. I. Neumann, Friedmund. II. Title. III. Title:
Testicular descent. IV. Series.
QL801.E67 vol. 81 [QP255] 599'.016 83-6666
ISBN-13:978-3-642-69123-2 (U.S.)

Composition: Schreibsatz Service Weihrauch, Würzburg

2121/3321-543210

# Contents

# 1 Introduction: Male Sexual Differentiation

## 1.1 Development of the Androgen-Dependent Structures of the Genitals

It has been demonstrated in the course of the past few decades that the presence or absence of androgens plays a major role in the development of the bisexual genital anlage of the vertebrates in a male or female direction (Bouin and Ancel 1903; Keller and Tandler 1916; Keller 1922; Lillie 1916, 1917; Jost 1948, 1967; Neumann and Hamada 1963; Neumann and Kramer 1964).

It is now regarded as definite that the persistence of the Wolffian ducts and their differentiation into vasa deferentia, seminal vesicles, and ejaculatory ducts, differentiation of the external genitalia in a male direction, and determination of the mode of tonic luteinizing hormone (LH) secretion are androgen-dependent in mammals. Recently, however, the existence of a cerebral center for the cyclic and thus also for the tonic secretion of LH has been disputed — at least in primates. According to these latest findings, the decisive element in the cyclic secretion of LH is not a particular determination of the brain, but the ovarian production of estradiol (Knobil 1980).

In contrast, determination of the gonadal sex and — as a further process of somatic differentiation — regression of the Müllerian ducts are not androgen-dependent. According to Jost (1947, 1970), a second hormone of the fetal testes, the so-called Factor X or anti-Müllerian hormone (AMH) (Jost 1947; Josso 1972) is responsible for regression of the female duct system in the male embryo: Jost's elegant fetal rabbit castration experiments demonstrated that castration of either male or female fetuses resulted in the regression of the Wolffian ducts and development of the Müllerian ducts. Replacement of testosterone stimulated the Wolffian ducts, but the Müllerian ducts continued to develop autonomously. These experiments are the basis for his conclusion that another hormone than testosterone must be responsible for the regression of the Müllerian ducts. The source of secretion of AMH is thought to be the Sertoli cells of the fetal testis: only Sertoli cells are able to induce the regression of Müllerian ducts maintained in organ cultures (Blanchard and Josso 1974; Tran et al. 1977; Josso 1974 a, b).

In human beings the activity of AMH has been demonstrated to be present before any differentiation of the Leydig cells has taken place (Josso 1971, 1973). This also argues in favor of the Sertoli cells as the site of origin of AMH, because the development of the Sertoli cells precedes the development of the Leydig cells.

The macromolecular nature of AMH has largely been elucidated. It is known to be a glycoprotein with an atomic weight of between 120 000 and 200 000 (Tran et al. 1977; Josso et al. 1977; Picard et al. 1978; Josso and Briard 1980; Budzik et al. 1980).

## 1.2 Testicular Descent

### 1.2.1 Significance

Whether or not another important developmental process in most male mammals, testicular descent, is androgen-dependent is controversial. It is defined as the descent of the male gonads from the abdominal into the scrotal position, i.e., movement from inside the body to the "world outside". The biological purpose of this phenomenon has been the subject of lengthy and continuing discussion, but no answer is as yet forthcoming (Starck 1975; Portman 1948).

It is frequently argued that the temperature inside the body is too high for spermatogenesis, which is why the testes descend from the abdominal cavity (Moore and Oslund 1924; More 1951). This argument, however, can be countered by the fact that there is no descent at all in some mammalian species (Monotremata, some insectivores, Edentata, Proboscidea, Cetacea), and that this has no adverse effect on the fertility of these animals. On the other hand, since even a slight increase in temperature results in considerable disturbances of spermatogenesis in mammalian species which naturally display testicular descent (Nelson 1951), it can be assumed that the sensitivity to temperature represents adaptation rather than the reason for descent.

If the reason for testicular descent really were sensitivity to temperature, it would also be extremely difficult from an evolutionary point of view to understand how the particular mammalian species that do not display descent could survive under these circumstances.

Attempts to establish a correlation (mode of life, biotope etc.) between those groups of animals which display descent and those in which the testes remain permanently in the abdominal cavity have likewise failed to clarify the question of the significance of this process. The only significant finding to emerge from these studies is that the main body temperature in those animal species without testicular descent is lower than in those with testicular descent (Wislocki 1933). Portman (1948), however, points out that the animals which do not display descent are — with the exception of the whale (Cetacea) — very old species in evolutionary terms, and suggests a relationship between the emphasis of sexual pole (bright coloring of the anal and sexual regions, development of hair) and testicular descent. In his opinion, features such as hair coloration, development of antlers, and the emphasis of the scrotal region just mentioned are all on a par with each other and serve the dual purpose of sex announcement and rank orientation.

### 1.2.2 Morphological Principles

The bisexual gonadal primordium of mammals develops in close association with the mesonephros at the posterior wall of the coelom (Romer 1971; Torrey 1945; Witschi 1951). The mesonephros and the gonads are connected to each other by a common peritoneal fold, the urogenital folds (plica urogenitalis). Cranial and caudal suspensory ligaments develop in both sexes from the cranial and caudal section of the mesonephros, which do not reach full maturity. The continuation of the caudal suspensory ligament corresponds to the urogenital cord with Müllerian and Wolffian ducts. Again in both sexes, the caudal mesonephric ligament becomes detached from the urogenital cord in the course of ontogenesis, and makes contact with the anterior abdominal wall (Starck 1975), after which it is known as the inguinal ligament of the mesonephros or gonad. In the male, the gubernaculum testis develops from this inguinal ligament (Starck 1975; Josso 1977; Gier and Marion 1969).

According to Starck (1975), the gubernaculum itself consists of two different sections, namely a free section and a section which stands in close association with the abdominal wall. The first section runs as the pars abdominalis gubernaculi from the mesonephros (the future efferent ductules of the epididymis) to the abdominal wall,

2

while the second passes through the abdominal wall as the pars inguinalis gubernaculi. The pars inguinalis gubernaculi and the abdominal wall muscles develop synchronously, so that the developing gubernaculum is surrounded by the developing abdominal wall muscles, the inguinal cone thus being formed. The gubernaculum therefore does not perforate the abdominal muscles. Once the inguinal cone has formed, evagination takes place bilaterally in this region, giving rise to the processus vaginalis peritonei. In the course of further development the entire inguinal cone everts into the scrotum to form the cremaster sac.

This process applies in principle to all mammals with testicular descent, although there are considerable species-specific differences as regards the details. In man and in many other mammals, for example, the processus vaginalis becomes obliterated after descent is complete, while it remains open in some other species. In the latter case, communication with the abdominal cavity is maintained permanently. This applies, for instance, to ungulates, in which the testes are present in the scrotum only during the rutting season. Communication with the abdominal cavity also persists in rodents, although the testes are normally always in the scrotum (Josso 1977; Gier and Marion 1969; Wensing 1968). The time at which the processus vaginalis is formed and descent is concluded also varies between the different species. Provided that development is undisturbed, the testes are fully descended in man at the time of birth (Lipshultz 1976; Scorer 1964; Starck 1975), whereas in rodents the processus vaginalis is not formed until after birth and descent is only completed during the first few postnatal weeks (Wensing 1968; Josso 1977).

In the rat, the testes are situated lateral to and at the same level as the kidneys on the 16th day post coitum. During the next 3 days the testes migrate to the inner inguinal ring, a distance of about 2–2½ mm. At the time of birth the testes are in immediate contact with the inner inguinal ring (Wensing 1968).


## 1.2.3 Processes

However, how testicular descent takes place on the basis of the morphological situation described — quite apart from its regulatory mechanisms — is not fully clear.

According to Gier and Marion (1969, 1970), the process of testicular descent can be divided into three different stages. The first phase consists of the nephric displacement and is closely related to the degeneration of the mesonephros. There is a rapid growth and anterior migration of the metanephros simultaneously with the degeneration of the mesanephros. As the mesonephros degenerates from anterior to posterior, the metanephros grows and migrates anteriorly, dorsal to the mesonephros, at least lying in an anterior position to the mesonephros. Further increasing in size, the metanephros forces the mesonephric remnants and the gonads further caudally.

It is suggested by Gier and Marion (1969, 1970) that the increase in coelomic fluid pressure is the decisive factor for this stage of descent. The increase in pressure is caused by contraction of the umbilical ring and the resultant separation of the embryonal and extraembryonal coelom, the processus vaginalis thus being formed as a hernia in the weak triangle.

The increasing abdominal pressure induces the processus vaginalis to protrude against a resistant gubernaculum. Because the fibrous gubernaculum elongates less rapidly than the processus vaginalis, the tension is transmitted to the Wolffian ducts and testes, causing the testes to be pulled posteriorly into the inguinal canal. During the third phase, called the "inguinal passage", the tension of the gubernacula and the pressure of the abdominal organs forces the inguinal cone to open, thus allowing the testes to slip through the everting inguinal cone into the scrotum. To summarize the findings of this group, the decisive event for testicular descent is the increase in coelomig pressure and its transmission to the testes via the gubernacula.

According to Wensing (1968, 1973 a, b) and Wensing and Colenbrander (1973), however, the decisive factor for testicular descent is a so-called swelling reaction of the gubernaculum. They compare this process with the reaction of a pear-shaped balloon which is inflated while lying partly within and partly outside a constriction. What happens is that the free part is pressed against the edge of the constriction, while the captive part is gradually drawn into the inflated free part. Wensing equates the captive part of the balloon with the intraabdominal and intrainguinal portion of the gubernaculum, and the free part with the extraabdominal portion. The extraabdominal portion of the gubernaculum is regarded as that part which, at the time of the swelling reaction, already lies distal to the inguinal canal. In the course of the swelling reaction, traction is exerted on the testes, thereby drawing the gonads into the inguinal canal. During a second phase postulated by Wensing, signs of regression then occur at the gubernaculum, which result in further traction on the testes and bring about conclusion of descent (Wensing 1968; Wensing and Colenbrander 1973). This theory stands only if the gubernaculum is not a fibromuscular band — a confirmed fact according to Wensing, but not according to many other authors (Gier and Marion 1969; Josso 1977; Bergin et al. 1970).

Other factors involved in testicular descent include contraction processes of the gubernaculum or of the skeletal muscles of the inguinal cone (Prasad 1974).

### 1.2.4 Regulatory Mechanism

The morphological process of testicular descent is controversial enough, but there is even less agreement regarding its regulatory mechanisms. The following are just a few examples from the plethora of literature available on this subject.

#### 1.2.4.1 Androgens, Human Chorionic Gonadotropin, and Luteinizing-Hormone-Releasing Hormone

Raynaud inhibited testicular descent by treating gestating mice with estrogens (Raynaud 1940, 1942, 1957, 1958). He interpreted this result as an antiandrogenic effect of the estrogens, concluding that testicular descent is androgen-dependent. By administering high doses of androgens to gestating rats, Greene achieved caudal displacement of the gonads in the female fetuses (Greene et al. 1939). Hadziselimovic also believes in androgen dependence on the basis of his studies on the testes of babies with cryptorchidism and on the basis of experiments with estrogens in gestating mice (Hadziselimovic 1977; Hadziselimovic and Herzog 1976, 1977). In the latter experiments he treated gestating mice with a single dose of estrogen on the 14th day post coitum, achieving by the time of birth (between days 17 and 19) the same degenerative changes of the Leydig cells which he also observed in cryptorchid babies aged 12 months. By treating the mice simultaneously with human chorionic gonadotropin

(HCG) he prevented the signs of atrophy of the Leydig cells in the male fetuses. Hadziselimovic interpreted these results as primary effects, claiming that incomplete descent is caused by deficient LH secretion, which in turn results in reduced testosterone production.

The more or less successful treatment of boys with cryptorchidism by means of HCG or LH-RH (luteinizing-hormone-releasing hormone) is also frequently used as an argument for the androgen dependence of testicular descent (Happ et al. 1975, 1977; Kollman et al. 1977; Hadziselimovic et al. 1977; Prader et al. 1976; Waaler 1979). However, the treatment of maldescended testes with LH-RH or HCG is not based on a full understanding of the regulatory mechanisms of testicular descent involving androgen dependence, success resulting from this method. Reports on success vary between 0 and 100% (Knorr et al. 1977).

One of the main problems in interpreting and explaining these very conflicting results is the precise definition of the type of cryptorchidism being reported (Lipshultz 1976; Scorer 1964). According to Scorer and Farrington (1971), disturbances of testicular descent comprise two main groups: (a) retractile and (b) truly maldescended testes. The latter group can be further categorized into functionally dystopic (i.e., high scrotal or inguinal position); abdominal; ectopic (lying outside the route of normal testicular descent, i.e., in the perineum, in the femoral region, or in front of the symphysis pubis); or obstructed (i.e., the normal way being closed).

The retractile testis, which is due to a hyperactive cremaster reflex, will normally descend with maturation (Lipshultz 1976). This type of maldescended testis accounts for 70% of cases and cannot be considered true cryptorchidism. The other 30% is accounted for by the second group (truly maldescended).

According to Scorer (1964), the distribution of testicular maldescent is as follows: ectopic, less than 1%; obstructed, 23%; inguinal, 19%; high scrotal, 49%; and abdominal or congenitally absent, 9%. Thus the high scrotal type is the most frequent one encountered within the group of truly maldescended testes.

Confusing data on the chances of spontaneous descent are also due in large part to unclear definition of the kind of cryptorchidism reported. Scorer and Farrington (1971) suggested that a spontaneous descent was possible only until the 9th month of life, if at all, and it is unlikely to occur after the 1st year of life unless it is actually a retractile type of testicular maldescent.

When the successful treatment of cryptorchidism by means of HCG or LH-RH is seen in this light, it becomes apparent that these hormones are most effective in cases of retractile testes. Given an adequate hormone therapy, Bierich (1977) postulated a change of 50% to get the testes descended. However, looking at the type of cryptorchidism being successfully treated, the retractile type of maldescended test's accounts for 92% whereas the so-called true cryptorchidism accounts for only 8.5%. Bierich himself found that the abdominal and fixed inguinal testicles are particularly resistant to the hormone therapy (HCG, LH-RH). These findings are in agreement with those reported by Knorr et al. (1977).

### 1.2.4.2 Antiandrogens

Experiments with the antiandrogen cyproterone acetate have produced discordant results. Treatment of the mother animals with cyproterone acetate led to inhibition of testicular descent in some studies (Neumann and Kramer 1964; Neumann et al.

1965), while the same treatment had no effect in others (Richter 1973; Elger et al. 1971, 1977).

Wensing, who initially believed that his "swelling reaction" of the gubernacula was comparable to the cockscomb reaction and therefore androgen-dependent, later revised his opinion after experiments with cyproterone acetate: the swelling reaction of the gubernacula could be influenced neither by testosterone propionate nor by cyproterone acetate (Wensing 1973a; Wensing and Colenbrander 1977). Furthermore, in histochemical studies he was unable to demonstrate any $3\beta$- or $17\alpha$-$OH$ dehydrogenase activity in the gubernacula of rats during the time in question (days 16–20 post coitum), whereas the same studies produced a positive result in the interstitial cells of the testes during the same period of time (Wensing 1973a). An attempt to induce the swelling reaction of the gubernacula by means of androgens in female pigs also failed. However, studies of testicular feminization in pigs are interesting in this connection (Wensing et al. 1975). This syndrome, which is also known in man, constitutes androgen insensitivity due to an enzyme defect for the development of the androgen receptor, but it may also be based on inadequate reduction of testosterone to dihydrotestosterone, so that although androgens are formed they cannot act on the target organ (Wilson 1975; Mühlenstedt and Schneider 1979; Odell and Swerdloff 1976).

Complete reduction of the Wolffian ducts and their derivatives and the development of a small vagina and female external genitalia have been observed in these animals, which are males in terms of both genetic and gonadal sex. At the same time, however, the gubernacula were developed in a typically male manner and a normal swelling reaction had taken place.

### 1.2.4.3 Anti-Müllerian Hormone

These latter findings suggest that testicular descent is not androgen-dependent, although it does appear to be gonad-dependent: caudal displacement of the testes fails to take place when male mammals are castrated at the critical point in time (Raynaud 1957; Raynaud and Frilley 1947). In contrast, partial descent of the ovaries has been observed in the freemartin (Keller 1922; Wolff 1962).

The inhibition of testicular descent induced by treating gestating mice with high doses of estradiol was accompanied by partial persistence of the Müllerian ducts (Raynaud 1940, 1942, 1958). These findings were subsequently confirmed by other groups (Neumann et al. 1969; Josso and Tran 1979). This effect is not only seen when induced experimentally but has also been observed to occur naturally: Josso and Tran (1979) reported these condition in a bull in which the testes were in a pseudoovarian position. This situation has been described in man (Josso et al. 1977).

### 1.2.5 Endocrinology of Cryptorchidism

Great controversy exists even with respect to the hormonal status of cryptorchidism. Investigation of steroidogenesis, i.e., the role of the $\Delta^5$-pathway and $\Delta^4$-pathway in cryptorchid and normal testes of man respectively, has shown the $\Delta^5$-pathway to be the major pathway both in cryptorchid and in normal testicular tissue (Gupta and Rager 1977). In the study performed by Gupta and Rager, however, the absolute amount of conversion to testosterone was demonstrated to be lower in cryptorchid children than in normal ones. Nevertheless, increased, decreased, and normal concentrations of testosterone have all been described. This also accounts for the concentration of LH and follicle-stimulating hormone (FSH) (Waaler 1979; Lee 1974; Hadziselimovic et al. 1977; Gupta and Rager 1977; Forest 1977).

Although the results of evaluation of hormone concentration in cryptorchidism comprise nearly all values imaginable, an intrinsic testicular abnormality or pathological process in the hypothalamic-hypophyseal-gonadal axis has never been truly substantiated (Lipshultz 1976; Lee et al. 1974). However, there is agreement by most authors concerning increasing concentrations of FSH with time, possibly as a result of progressive damage to the germinal epithelium (Lee et al. 1974; Hedinger 1979; Lipshultz 1976). But although the effects of hormones on cryptorchidism are under investigation, there is no proof as regards the genesis of cryptorchidism because it is, after all, impossible to differentiate between primary and secondary events. It therefore remains to be elucidated whether changes in the hormone profile are intrinsic ones causing cryptorchidism, which results in progressive degeneration of the germinal epithelium because of the abnormal position of the testes, or vice versa.

### 1.2.6 Experimental Studies

Experimental cryptorchidism is commonly used study testicular function under these conditions in the hope of shedding some light on the regulatory mechanisms of testicular descent. Artificial cryptorchidism is mainly achieved in studies using rats by drawing normally descended testes through the inguinal canal back to the abdomen and suturing them to the abdominal wall. Such investigations have been performed extensively and with results almost as conflicting as in the case of man.

For example, it has been reported in one study that serum gonadotropin levels were increased in the cryptorchid rat, and in another that they were decreased (Swerdloff et al. 1971; Keel and Abney 1980; Jones et al. 1977; Bergh and Damber 1979). Serum testosterone levels have been found to be reduced or unchanged (Gupta et al. 1975; Jones et al. 1977; Jean et al. 1975; Gupta 1979; Damber et al. 1978). Similarly different results have also been obtained for serum estradiol levels (Keel and Abney 1980; Jones et al. 1977; Hall and Gomes 1975).

Androgen-binding protein (ABP) secretion has been shown to be severely impaired in experimental cryptorchidism in the rat (Hanson et al. 1975; Ritzén et al. 1977), and Gupta (1979) observed differences in the pattern of in vitro steroidogenesis in intact and cryptorchid animals. Frowein and Engels (1975) found an enhacement in HCG-binding capacity in cryptorchid rat testes as compared to normal ones.

However, the main problem involved in the interpretation of the results of the experiments outlined above is the method itself. In experimental cryptorchidism testes which have originally descended in a normal way are replaced in the abdomen, so that it is impossible to decide whether the effects induced by that method have any physiological significance for the genesis of cryptorchidism or the regulatory mechanisms of descent. As a consequence, this method does not offer a system appropriate for studies on the regulatory mechanisms of descent.

The object of the present paper is to obtain further information about the cause of disturbances of descent with the aid of specific hormone treatments in gestating

7

rats at the time of sexual differentiation of the fetuses, and to shed more light on the regulatory mechanism of testicular descent. In this connection an attempt is also made to clarify the extent to which changes in the proportions of the genital system contribute to disturbances of descent.

# 2 Materials and Methods

The studies were conducted in Wistar rats — both in fetuses and in adult animals whose mothers had been treated. Female animals in estrus were placed with bucks over night. If spermatozoa were demonstrable in the vaginal smear on the following day, this day was counted as day 1 of pregnancy (1st day post coitum). All the fetuses were removed by cesarean section on day 22 post coitum. The studies refer only to the male animals. Significance was calculated with the aid of the Dunnet and Scheffe tests.

The gestating animals were treated subcutaneously with ethinyl estradiol, cyproterone acetate, or a combination of ethinyl estradiol and cyproterone acetate from day 17 to day 20 of pregnancy. The controls received only the solvent. Some of the fetuses removed on day 22 were immediately killed and fixed, while the others were wet-nursed by other female rats (the maternal animals themselves could neither give birth nor lactate because of the hormone treatment).

The wet-nursed animals were killed between 2 and 2½ months after birth. The exact number of examined animals, the subdivision of the groups, and the dosages used are presented in Tables 1 and 2. The tables contain only data on the litters from which animals were used for examination. The animals in which the postnatal development of the external genitals was studied are not included. The numbers of these animals are indicated of the respective chapters. The actual number of litters and mated animals was considerably higher — the marked abortifacient effect of the estrogens at the time of treatment in this species meant that a larger number of animals was required to obtain sufficient animal material for the studies. This abortifacient effect is also the reason for different doses of estrogens. The high dose (0.25 or 0.20 mg per animal) was administered only to those whose litters were to be killed

Table 1. Division and treatment of the adult male animals the mothers of which were treated

| Group | Number of litters | Dose | Number of adult male animals examined | |
|---|---|---|---|---|
| | | | Qualitative | Quantitative |
| I | 6 | | 8 | 7 |
| II | 6 | CPA 10 mg/d s.c. | 8 | 7 |
| III | 5 | EE 0.03 md/d s.c. | 6 | 6 |
| IV | 12 | EE + CPA 0.03 + 10 mg/d s.c. | 15 | 9 |

CPA, cyproterone acetate; EE, ethinyl estradiol

Table 2. Division and treatment of the fetuses

| Group | Number of litters | Dose (mg/d) | Number of fetuses examined |
|---|---|---|---|
| Ia | 5 | | 7 |
| IIa | 5 | CPA 10 mg/d s.c. | 7 |
| IIIa$_1$ | 2 | EE 0.25 mg/d s.c. | 2 |
| IIIa$_2$ | 3 | EE 0.20 mg/d s.c. | 5 |
| IVa | 7 | EE + CPA 0.25 + 10 mg/d s.c. | 8 |

CPA, cyproterone acetate; EE, ethinyl estradiol

immediately after birth. All the animals whose young were to be wet-nursed received only 0.03 mg per animal day.

The development of the external genitals and the internal position, shape, and size of the gonads of all male young were examined macroscopically during the first 56—85 days of life. The animals were decapitated and fixed, and the length of the genital tract (cranial pole of the gonad to phallus tip) and the length of the Wolffian ducts were determined macroscopically. The measurements were made with the aid of a slide rule.

The fetuses were transected at the level of the kidneys, fixed in Bouin, embedded in hard paraffin (56°) and then cut in transverse serial sections with a microtome (Jung). The sections were 4 mm thick, and every tenth section was mounted and stained with hematoxylin-eosin. Assessment of the fetuses was almost exclusively quantitative. The following were determined by counting sections:

The level of the genital folds (cranial gonadal pole to the junction in a middle ligament)

The level of the genital cord (junction of the gonaducts in a middle ligament to the opening of the gonaducts into the urethra

The level of the genital tract (cranial gonadal pole to urogenital sinus)

The length of the Wolffian ducts from the cranial pole of the genital cord up to the junction of the seminal vesicles and vasa deferentia with the ejaculatory ducts

The length of the ejaculatory ducts

The level of testicular descent (caudal renal pole to cranial testicular pole).

The degree of the development of the gubernacula and inguinal cone was determined qualitatively.

The genital tract of each wet-nursed animal was prepared in toto, attached to wax plates to prevent contraction of the organs, and fixed in Bouin. The region of the colliculus seminalis was cut in serial sections of 4 mm and every tenth section was mounted. The sections were stained partly with hematoxylin-eosin and partly with Heidenhain's azan stain.

Qualitative as well as quantitative evaluation of the sections was then performed. The length of the Wolffian duct (vasa deferentia) in the regions which could not be measured macroscopically was determined by counting the sections, as were also the length of the ejaculatory ducts and the length of the free urethra from the bladder neck to the formation of the urogenital sinus.

# 3 Results

## 3.1 Controls

The development of the external genitals of the male animals from ten litters was followed over a maximal period of 70 days. Macroscopic qualitative studies were conducted in 17 adult male animals, while quantitative studies to determine the proportions of the genitals were conducted in seven males. Microscopic qualitative studies were performed in eight males. Qualitative and quantitative studies were also conducted in seven male fetuses.

### 3.1.1 Qualitative Macroscopic Examination of the Adult Animals

#### 3.1.1.1 Development of the External Genitals of the Males

*Day 1 (Day of Birth).* The males are clearly distinguishable from the females by virtue of the position of the urethra, which opens out at the tip of the phallus.

*Day 8.* The signs of the scrotal anlage can be seen in some animals.

*Day 14.* The scrotal anlage is clearly recognizable in all animals. The phallus is distinctly further developed than in the female animals.

*Days 18–10.* The testes of most animals are in the inguinal region.

*Days 21–24.* The testes are almost completely descended, but still lie somewhat high in the scrotum.

*Days 24–28.* After 28 days at the latest, the testes of all animals in this group have completely descended into a normally developed scrotum.

*Days 60–70.* The external genitals display no more major changes before the animals are killed.

#### 3.1.1.2 The Internal Genitals

The internal genitals of all animals examined display regular differentiation with no peculiarities. The testes are situated in a scrotum which communicates with the abdominal cavity. The testes are seperated from the scrotal epithelium by the parietal layers of the cremaster sac, to the distal portion of which the gubernacula are attached.

The other parts of the genital system also show normal differentiation.

### 3.1.2 Qualitative Microscopic Examination of the Adult Animals

Microscopically there is no deviation of the regular development. All parts of the genitals are normally differentiated (Figs. 1–3). The Müllerian ducts are regressed except for the normally present rudiment – the prostatic utricle (Fig. 4).

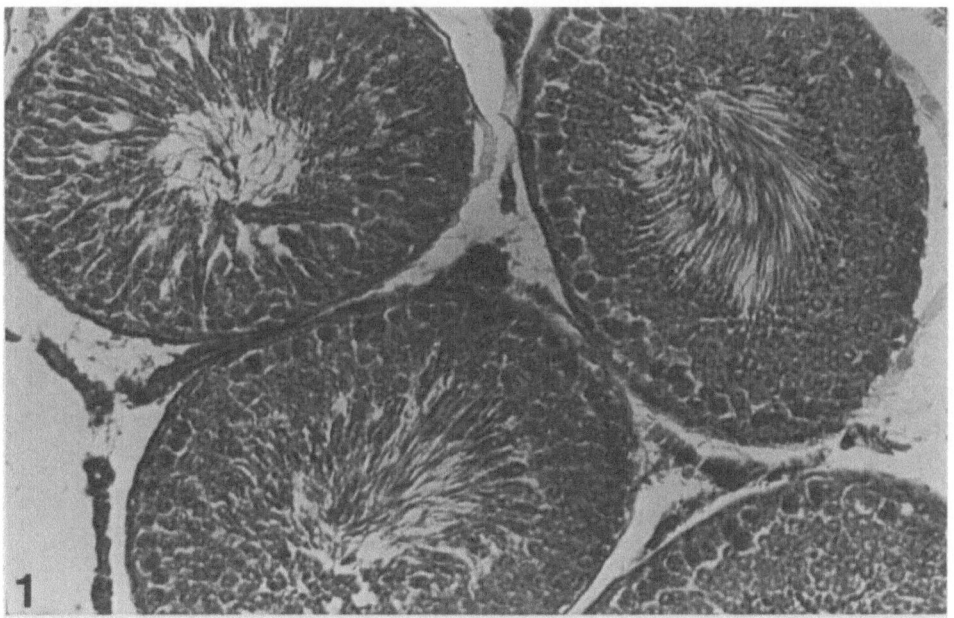

Fig. 1. Testis of an adult control animal. All stages of spermatogenesis up to spermatozoa are visible. × 320

### 3.1.3 Quantitative Microscopic and Macroscopic Examinations of the Adult Animals

The genital tract — from the cranial gonadal pole up to the tip of the phallus — achieves a mean length of 87.52 mm. The vasa deferentia account for an average of 52.55 mm, the ejaculatory ducts for 0.30 mm, and the urogenital sinus for 35.47 mm. The free section of the urethra, i.e., the section from the neck of the bladder up to the formation of the urogenital sinus, displays a mean length of 2.64 mm (Tabelle 3).

When the individual sections of the genitals are expressed as percentages of the entire length, the vasa deferentia are found to account for 60.0%, the ejaculatory ducts for 0.34%, and the urogenital sinus for 40.53% (Table 4, Figs. 5 and 6).

Table 3. The influence of cyproterone acetate, ethinyl estradiol, and ethinyl estradiol + cyproterone acetate, administered to gestating rats from day 17 to day 20, on the length of the genital tract and its individual sections in the male adult young (Values in Millimeters)

| | GT | VD | DE | Ufr | US |
|---|---|---|---|---|---|
| Controls | 87.52 ± 1.92 | 52.55 ± 3.56 | 0.3 ± 0.04 | 2.64 ± 0.25 | 35.47 ± 2.85 |
| CPA | (+) 71.84 ± 1.05 | (−) 48.48 ± 1.36 | (+) 0.93 ± 0.12 | (+) 5.57 ± 0.55 | (+) 22.06 ± 1.42 |
| EE | (+) 69.15 ± 1.45 | (−) 44.53 ± 1.45 | (+) 1.57 ± 0.13 | (+) 4.01 ± 0.19 | (+) 23.02 ± 0.34 |
| EE + CPA | (+) 72.18 ± 1.68 | (−) 50.77 ± 2.31 | (+) 1.32 ± 0.11 | (+) 5.42 ± 0.66 | (+) 20.02 ± 1.7 |

GT, genital tract; VD, vasa deferentia; DE, ejaculatory ducts; Ufr, free section of urethra; US, urogenital sinus; (+) indicates significant compared to the controls; (−) indicates not significant compared to the controls

11

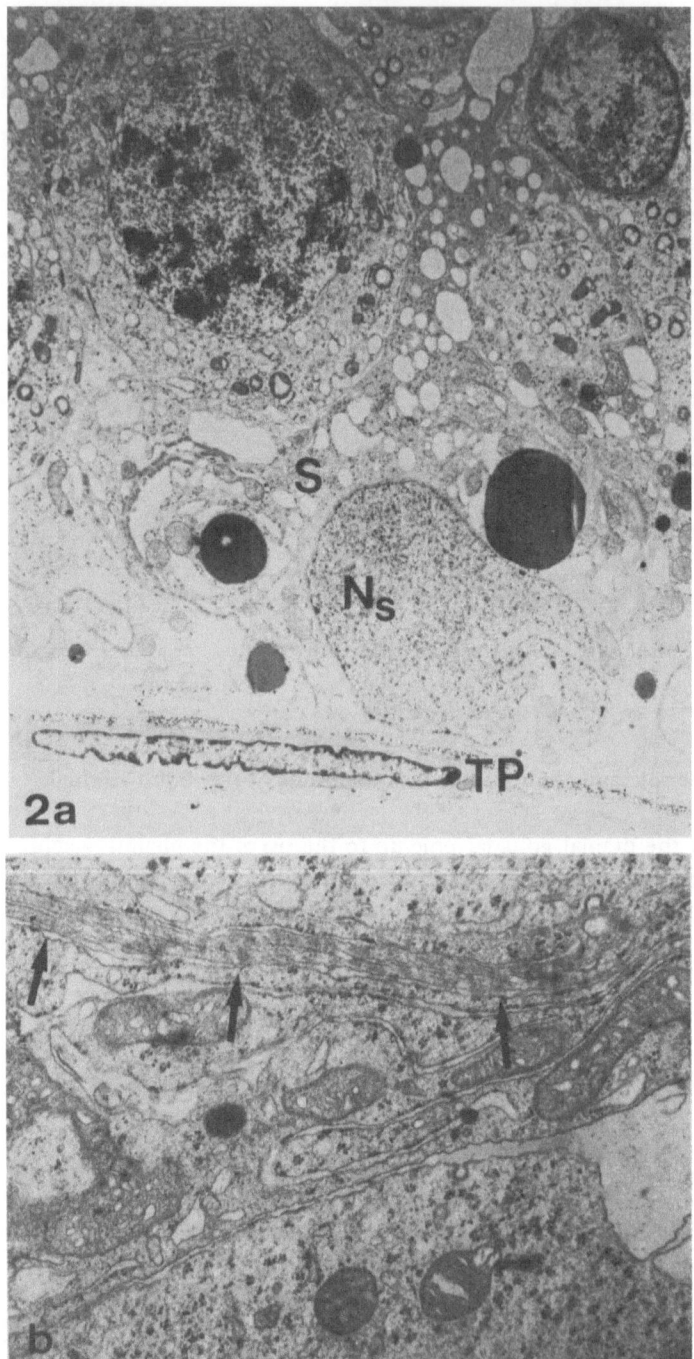

Fig. 2. *a* Basal compartment of a seminiferous tubule from a control animal. *S*, Sertoli cell; *Ns*, nucleus of Sertoli cell; *TP*, tunica propria. × 3300
*b* Blood-testis barrier from a control animal (*arrows*). × 23 700

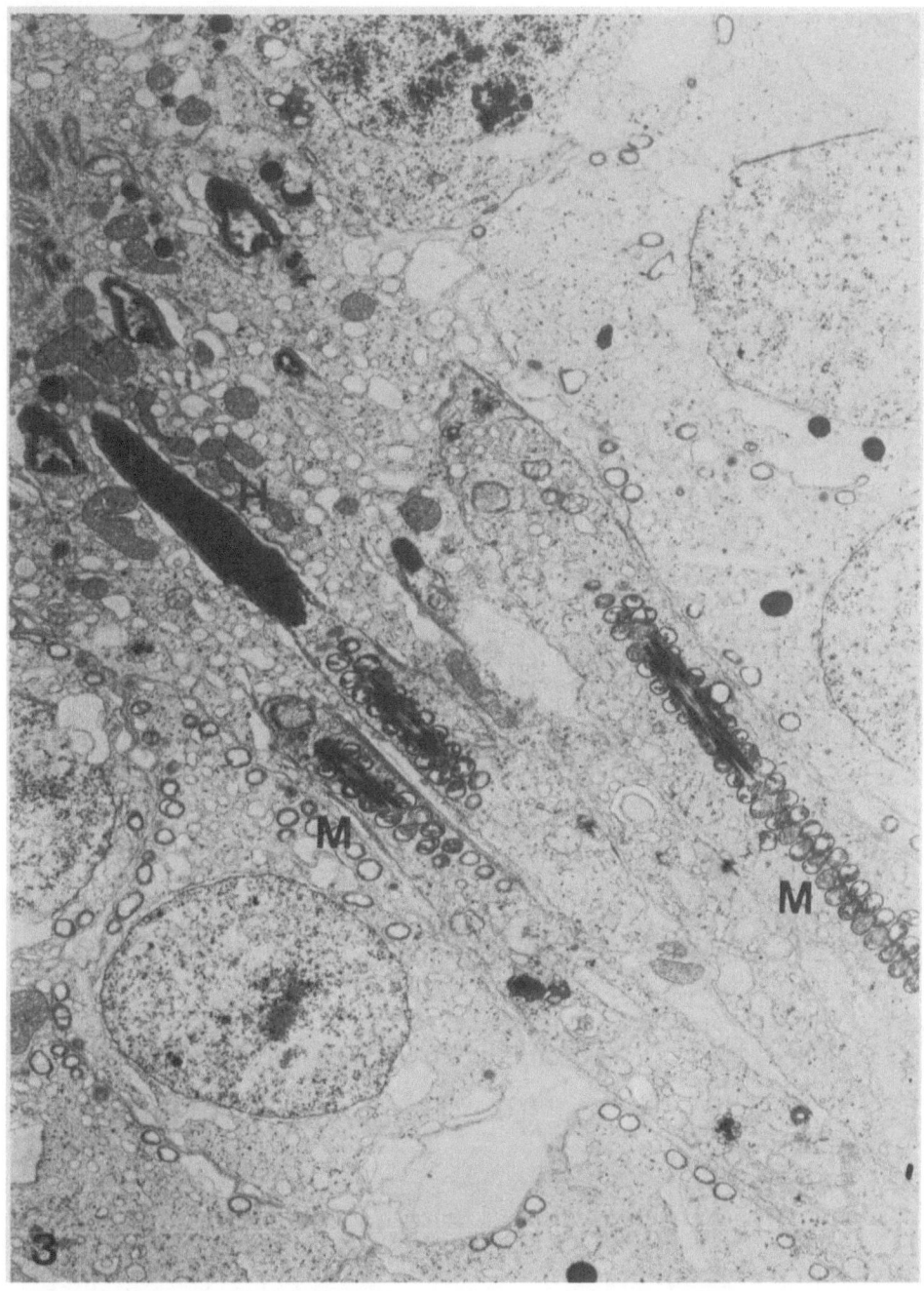

Fig. 3. Electron micrograph of normally developed spermatozoa from a control animal. *H*, head; *M*, mitochondria of the middle piece. × 5700

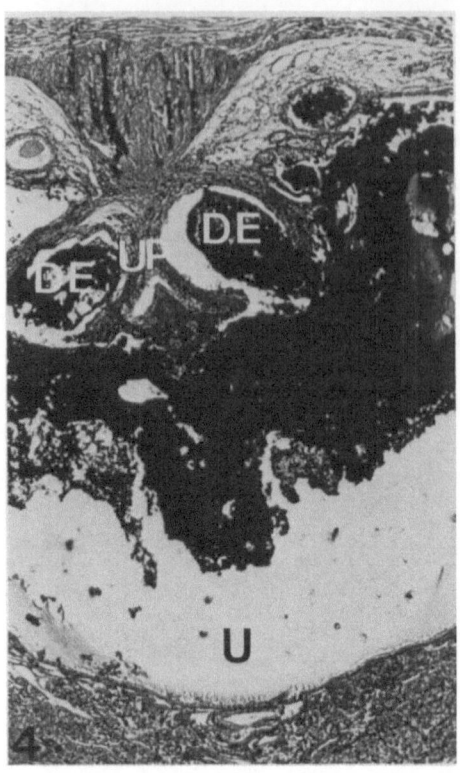

Fig. 4. Male control animal: formation of the urogenital sinus with the prostatic utricle (*UP*) in between. *DE*, ejaculatory ducts; *U*, urethra. × 64

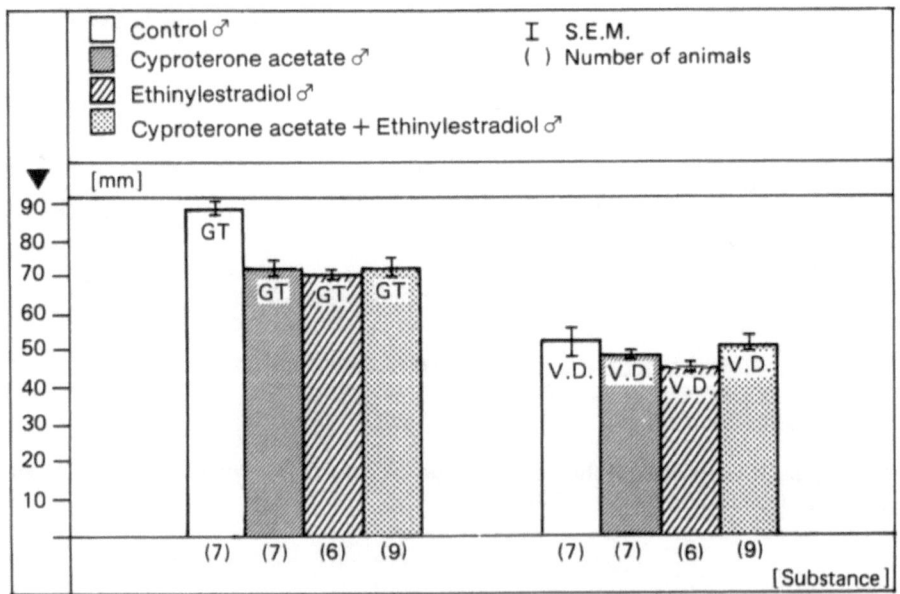

Fig. 5. The prenatal influence of cyproterone acetate, ethinyl estradiol, and ethinyl estradiol + cyproterone acetate on the length of the entire genital tract (*GT*) and the vasa deferentia (*VD*) of the male young

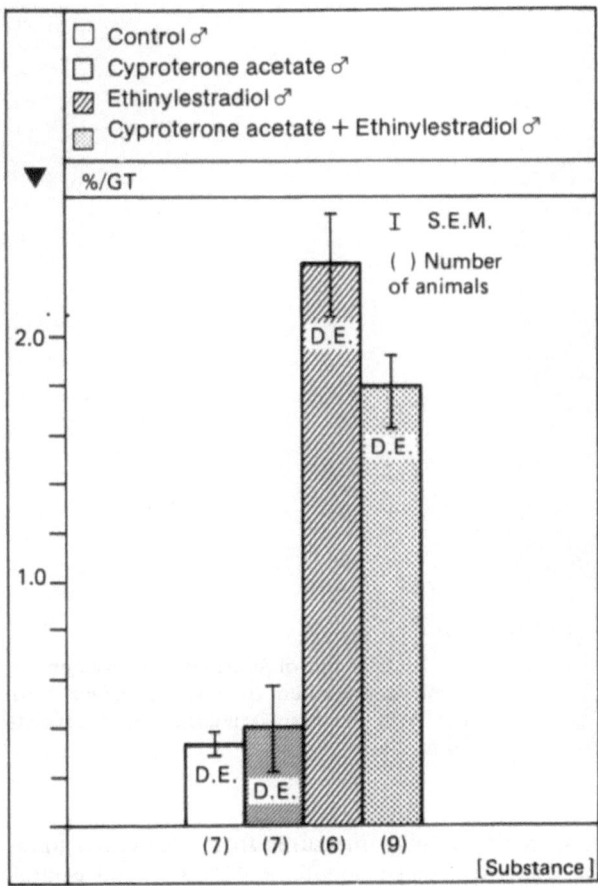

Fig. 6. The prenatal influence of cyproterone acetate, ethinyl estradiol, and ethinyl estradiol + cyproterone acetate on the length of the ejaculatory ducts (*DE*) as a proportion of the entire genital tract

Table 4. The relationship between the entire genital tract (GT) and the vasa deferentia (VD), the ejaculatory ducts (DE), and the urogenital sinus (US) in male adult rats the mothers of which were treated from day 17 to day 20 of pregnancy

|  | % GT | | | |
|---|---|---|---|---|
|  | Controls | CPA | EE | EE + CPA |
| VD | 60.04 ± 4.07 | 67.48 ± 1.89 | 64.4  ± 2.1 | 70.34 ± 3.2 |
| DE | 0.34 ± 0.05 | 0.42 ± 0.17 | 2.27 ± 0.19 | 1.83 ± 0.15 |
| US | 40.53 ± 3.26 | 30.71 ± 1.98 | 33.29 ± 0.49 | 27.74 ± 2.36 |

CPA, cyproterone acetate; EE, ethinyl estradiol

### 3.1.4 Male Fetuses

The testes have descended an average distance of 2.765 $\mu$m. The inguinal cones have developed normally, their muscles are strong, and the gubernacula are short and compact (Fig. 7). The testes are lying close against the neck of the bladder and are differ-

15

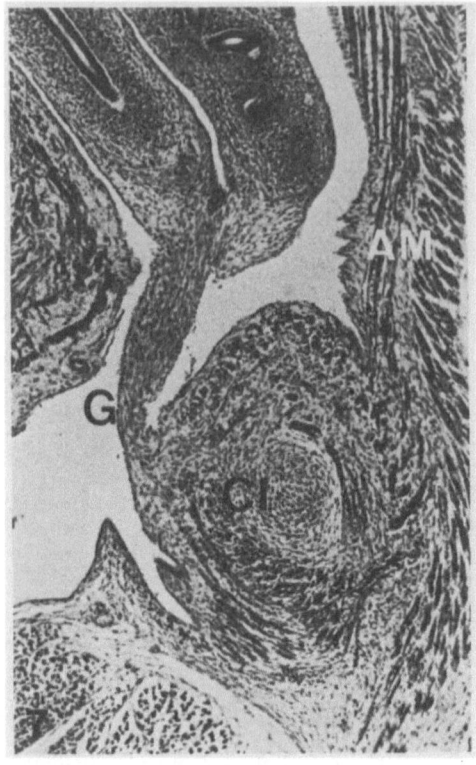

Fig. 7. Male control fetus: normal development of the gubernaculum (*G*) and the inguinal cone (*CI*). *AM*, abdominal muscles. Hematoxylin-eosin. × 64

entiated in keeping with the age. The genital tract — measured from the cranial gonadal pole to the urogenital sinus — achieves a mean length of 972 $\mu$m. The genital folds account for 470 $\mu$m of this, the genital cord for 665 $\mu$m. The length of the ejaculatory ducts is 182 $\mu$m, that of the vasa deferentia 462 $\mu$m (Fig. 8, Table 5).

## 3.2 Cyproterone Acetate

In the cyproterone acetate group (group II), the development of the external genitals of the male animals from ten litters was studied. Qualitative macroscopic examinations of the internal genitals were conducted in 21 adult animals. Quantitative examinations (microscopic as well as macroscopic) were conducted in seven adult animals. Qualitative microscopic examinations were performed in eight males; seven male fetuses were also examined, mainly from a quantitative point of view.

### 3.2.1 Macroscopic Examination of the Adult Animals

#### 3.2.2.1 Development of the External Genitals

*Days 1–11.* The animals cannot be sexed at the time of birth: no anlage of nipples can be recognized in any of the animals, and the phallus is split. Reliable sexing is possible after the 1st week — the anlage of nipples is just recognizable in the female animals on the 11th postnatal day, and the phallus is much further developed in the male animals than in the females.

Table 5. The influence of cyproterone acetate (CPA), ethinyl estradiol (EE), and ethinyl estradiol + cyproterone acetate on testicular descent and on the length of the Wolffian ducts (WG), the ejaculatory ducts (DE), the genital folds, the genital cord, and the entire genital tract of male rat fetuses the mothers of which were treated from day 17 to day 20 of pregnancy. Values in mikrometers ± SEM

| | CR-Cr. Te | WG-DE | DE | Genital folds | Genital cord | Genital tract | Number of fetuses |
|---|---|---|---|---|---|---|---|
| Controls | 2765 ± 175 | 462 ± 44 | 182 ± 2 | 470 ± 82 | 665 ± 45 | 972 ± 63 | 7 |
| CPA | 1772 ± 244 | 711 ± 16 | 223 ± 23 | 1118 ± 93 | 935 ± 32 | 2067 ± 112 | 7 |
| EE | 1289 ± 31 | 544 ± 31 | 317 ± 7 | 1261 ± 166 | 861 ± 44 | 2040 ± 186 | 7 |
| EE + CPA | 902 ± 147 | 719 ± 11 | 418 ± 22 | 1290 ± 97 | 1138 ± 56 | 2427 ± 122 | 8 |

CR-Cr. Te, caudal renal pole to cranial pole of the testes

## Further Development of the External Genitals of the Male Animals

The penis is not as well developed as in group I (controls) by the 11th day, but much further developed, than in groups III and IV (ethinyl estradiol and cyproterone acetate).

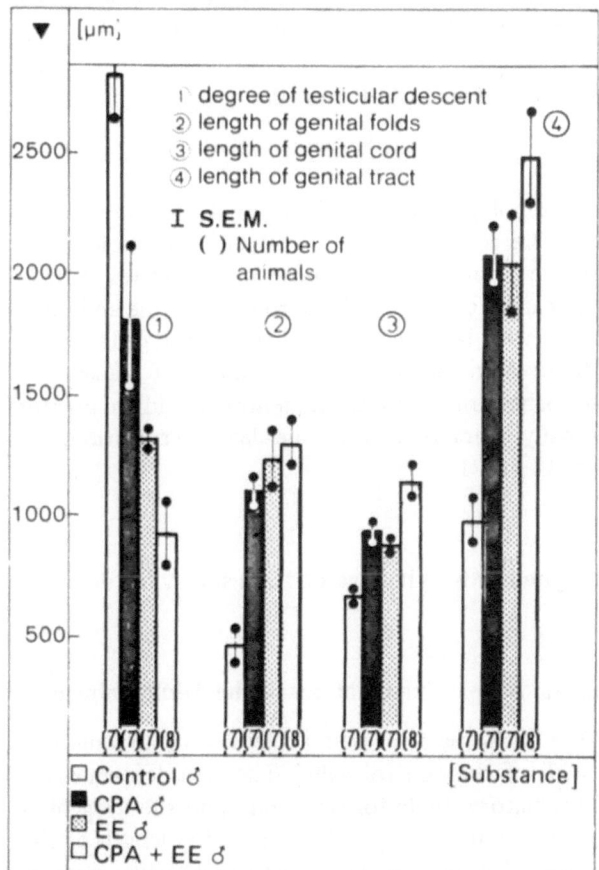

Fig. 8. The influence of prenatal hormone treatment on testicular descent and on the length of the genital folds, the genital cord, and the entire genital tract of rat fetuses. *CPA*, cyproterone acetate; *EE*, ethinyl estradiol

17

*Day 18.* The testes of almost all animals are in the inguinal region. Overall, the scrotum is not so well developed as in the control animals.

*Day 24.* The gonads of some animals have descended completely, some unilaterally, and some not at all. There is no scrotal anlage in these latter cases.

*Day 28.* After 28 days — the latest time at which descent was concluded in the control animals — the testes of 26 animals have descended into a scrotum which is distinctly less well developed than in the controls; unilateral descent, always of the right testis, has occurred in five animals; the gonads have descended into an extremely small scrotum in six animals; no descent at all has taken place in one animal.

*Day 32.* After another 4 days the testes of all animals except one have descended. The scrotal anlage still appears to be less developed than in the controls.

*Day 38.* The scrotum has achieved normal size in half of the animals in this group.

*Days 62–68.* By the day they are killed, all the animals display normally descended testes and a normally developed scrotum. The penis is distinctly less developed than in the control animals.

### 3.2.1.2 The Internal Genitals

The development of the internal duct system and the accessory sex glands is distinctly inhibited. However, the position, shape, and size of the testes are normal.

### 3.2.2 Qualitative Microscopic Examination of the Adult Animals

Testes

Under light microscopy, testicular morphology appears to be normal. No inhibition or disturbance of spermatogenesis is to be seen (Fig. 9). In principle, this picture is confirmed at the electron microscopic level (Figs. 10 and 11a). However, Sertoli cells with strong signs of high activity can be observed. The chromatin is fine and evenly distributed, as it is generally in cells with high activity like the Sertoli cells. Nevertheless, compared to the control there is an increase in the amount of the smooth endoplasmic reticulum. Also, the mitochondria are bigger than normal and are present more frequently. The nucleus, normally characterized by one deep invagination, has often a rather rounded shape (Figs. 11 and 12).

Persistence of Müllerian derivatives

A vagina persists in all animals investigated. It is to be regarded as a somewhat enlarged prostatic utricle (Figs. 13–15).

### 3.2.3 Quantitative Microscopic and Macroscopic Examinations of the Adult Animals

In this group the genital tract of the adult male animals achieves a mean length of 71.84 mm, which corresponds to 82% of the control value. The vasa deferentia account for 48.48 mm of this, the ejaculatory ducts for 0.93 mm, and the urogenital sinus for 22.06 mm. The most striking finding is the reduction of the length of the urogenital sinus to 62.22% of the normal value. The length of the Wolffian ducts is

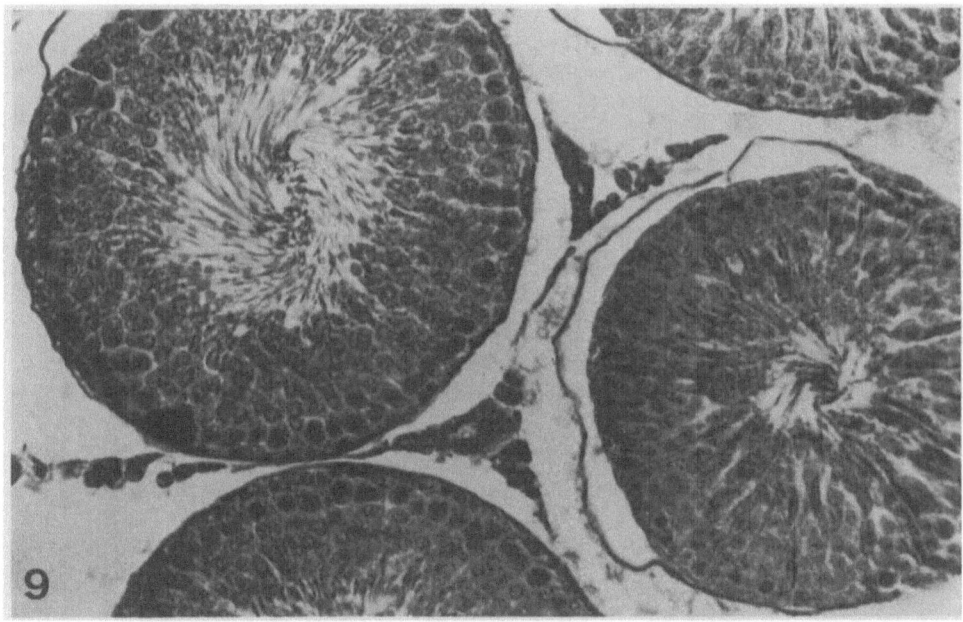

Fig. 9. Testis of a male rat the mother of which had been treated with 10 mg cyproterone acetate from day 17 to day 20 of pregnancy. At the light microscopic level no deviation can be seen in comparison to the control. Hematoxylin-eosin, × 320

also somewhat reduced, but not significantly in comparison to the controls. In contrast, the length of the ejaculatory ducts displays a three-fold increase. The length of the free urethra has also increased considerably to 5.57 mm or 210.98% of the normal value (Table 3, Fig. 5).

The proportions of the individual sections of the genital tract are as follows: the vasa deferentia account for 67.48%, the ejaculatory ducts for 0.42%, and the urogenital sinus for 30.71%. The correlations between the entire urethra, the free section, and the urogenital sinus also shift. The urogenital sinus accounts for only 79.18% of the entire urethra, while the free part increases to 20.16% (Figs. 6, 16 and 17, Table 4).

### 3.2.4 Male Fetuses

The testes have descended normally in only two animals. In the other five they are situated either with their caudal pole at the cranial end of the bladder, not close up to it, but far apart at the lateral wall (two animals) or dorsolateral from the cranial end of the bladder (three animals). The testes have descended a mean distance of 1.732 $\mu$m. In comparison, therefore, the testes of the control animals lie an average of 1.000 $\mu$m further caudally. The testes are longer and thinner than in the controls. However, their histology reveals no peculiarities. The gubernaculum and inguinal cone display normal differentiation in all fetuses (Table 5, Fig. 18).

The genital cord has a mean length of 995 $\mu$m, compared to 665 $\mu$m in the controls. Even more pronounced is the difference between the length of the genital folds in the animals of this group and those of the controls — 1118 $\mu$m compared

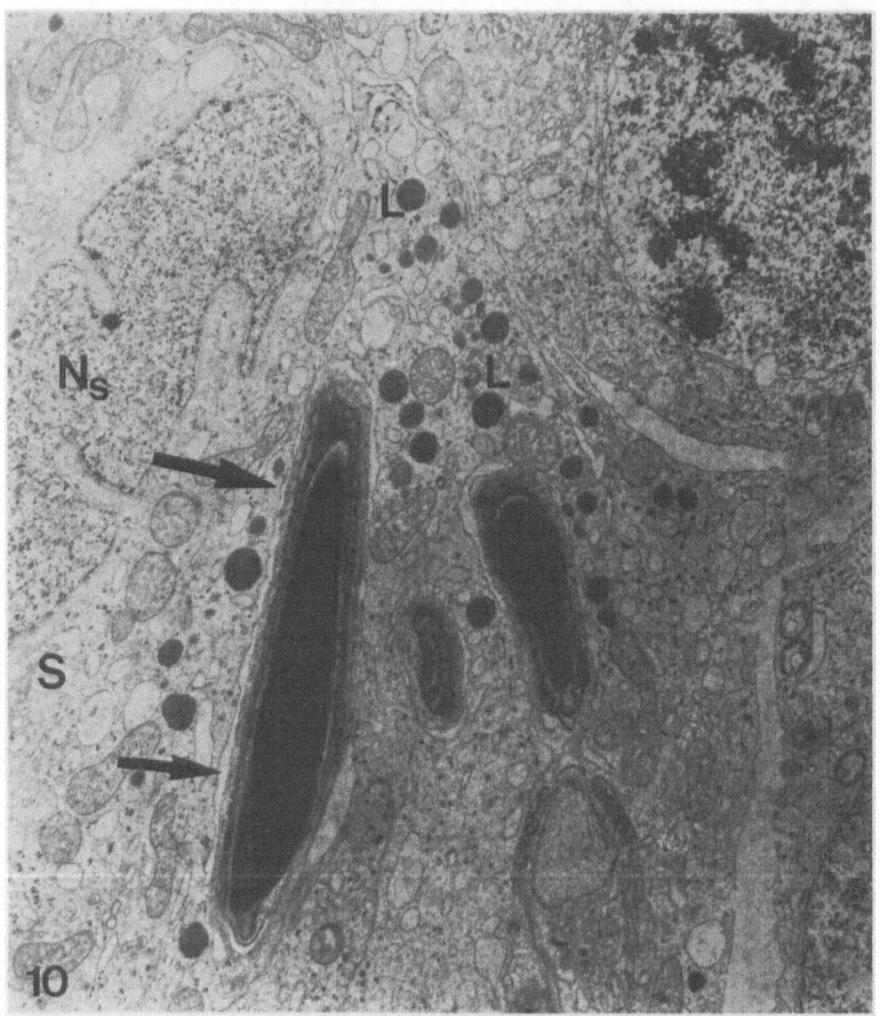

Fig. 10. Seminiferous tubule from a male rat the mother of which had been treated with 10 mg cyproterone acetate from day 17 to day 20 of pregnancy. Even at the electron microscopic level the heads of spermatids (*arrows*) are intact. Many lysosomes (*L*) are visible within the Sertoli cell (*S*). *Ns*, nucleus of Sertoli cell. × 11 200

to 470 μm – although even this does not approach the length of the genital fold in the female control animals (2584 μm). At 711 μm, the Wolffian ducts are considerably longer than in the controls, and the length of the ejaculatory ducts has also increased distinctly. The genital tract as a whole achieves a length of 2067 μm in the fetuses of this group, compared to a normal value of 972 μm (Table 5, Fig. 8).

## 3.3 Ethinyl Estradiol

In the ethinyl estradiol group (group III) the development of the external genitals of the male animals from five litters (nine male animals) was studied over a period

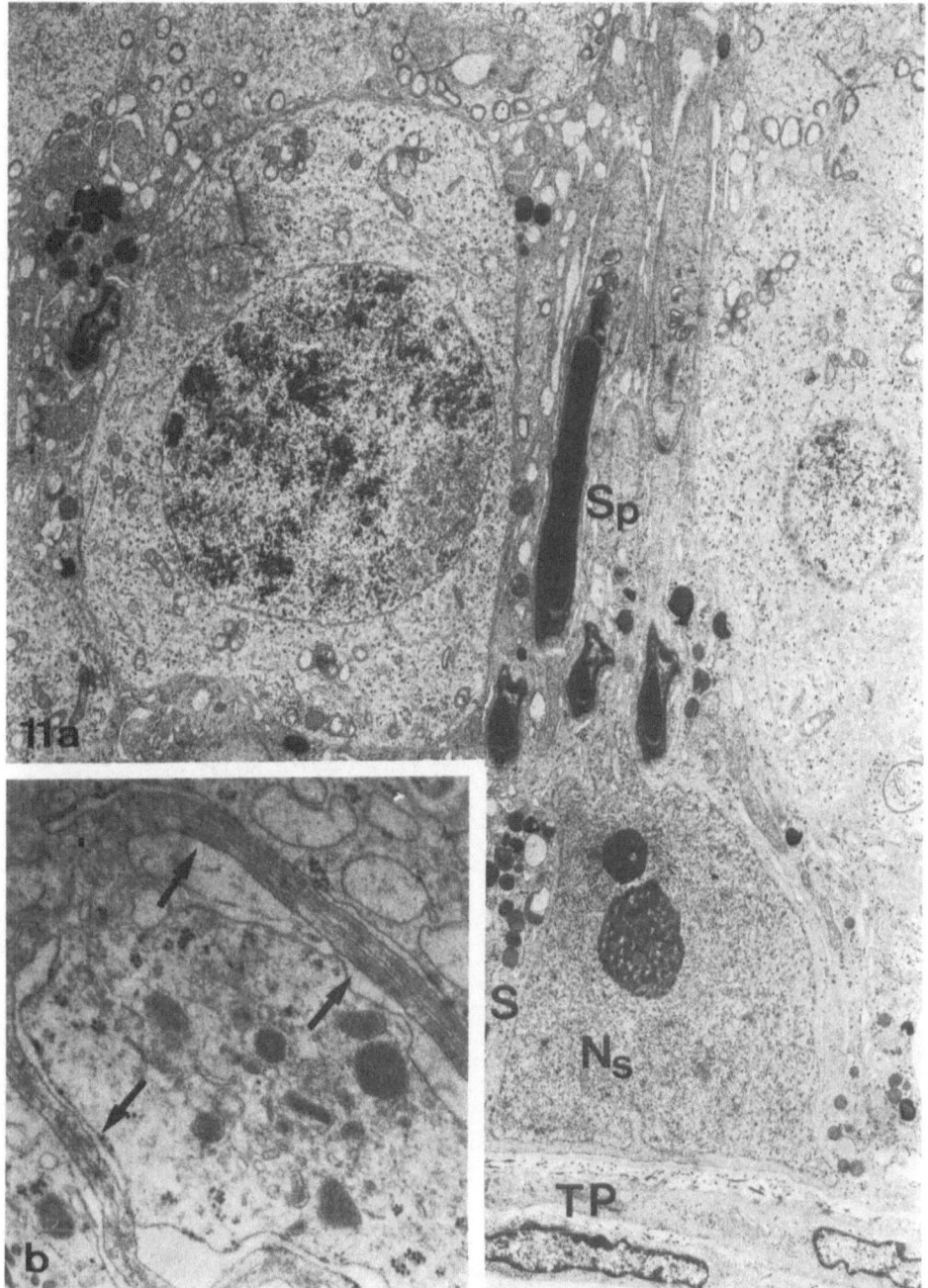

Fig. 11. *a* Seminiferous tubule from an adult animal the mother of which had been treated with 10 mg cyproterone acetate from day 17 to day 20 of pregnancy. The Sertoli cell is characterized by a poorly invaginated nucleus and a large nucleus. *S*, Sertoli cells; *Sp*, spermatid; *TP*, tunica propria; *NS*, nucleus of Sertoli cell. × 6650

*b* Normally developed blood-testis barrier (*arrows*) in an adult animal the mother of which had been treated with cyproterone acetate from day 17 to day 20 of pregnancy. × 28500

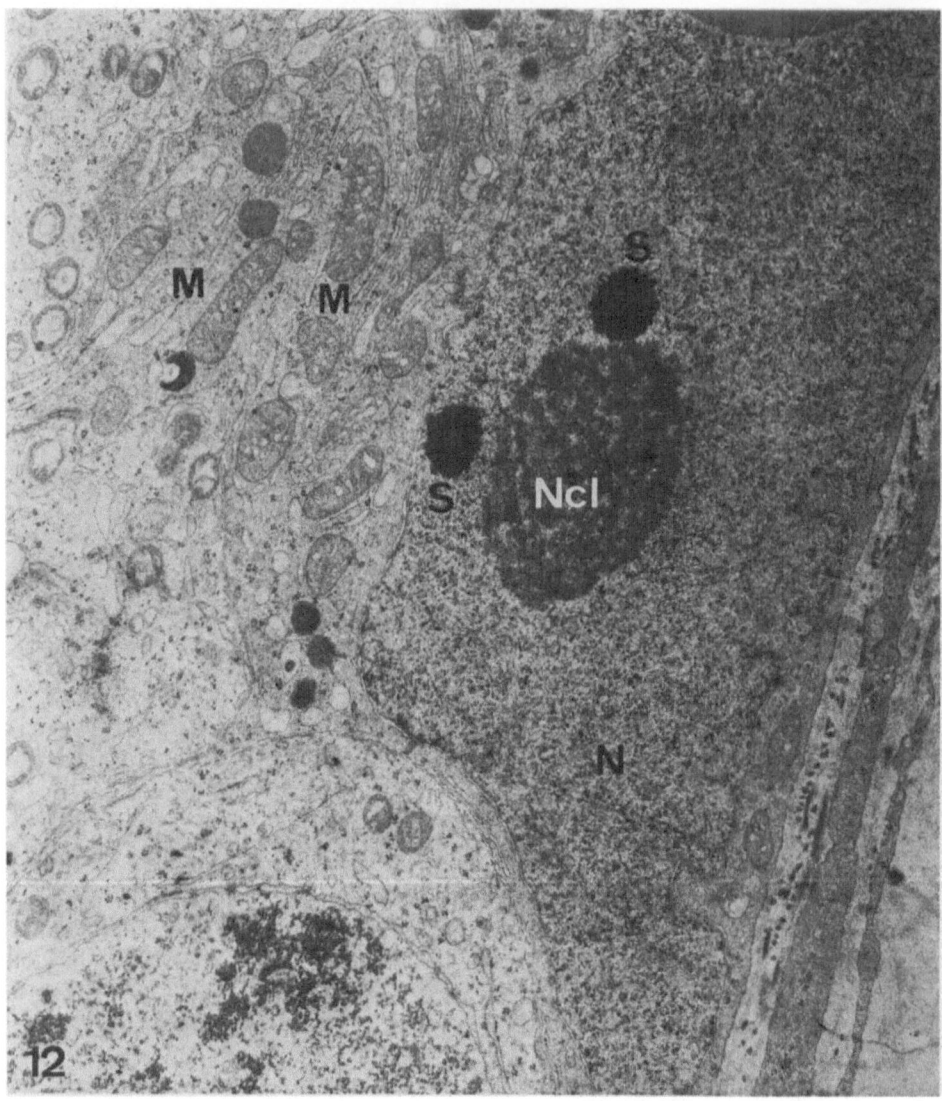

Fig. 12. Enlarged section of a highly active Sertoli cell. The mother of the animal had been treated with cyproterone acetate from day 17 to day 20 of pregnancy. *N*, nucleus; *Ncl*, nucleolus; *S*, satellite; *M*, mitochondria. × 13 200

of 56 days. Qualitative macroscopic studies of the internal genitals and quantitative studies (microscopic as well as macroscopic) to determine the proportions of the genital tract were each conducted in six adult male animals. Qualitative microscopic examinations were also performed in six males. Seven male fetuses were examined, mainly from a quantitative point of view. The development of the gubernacula and inguinal cones was studied qualitatively.

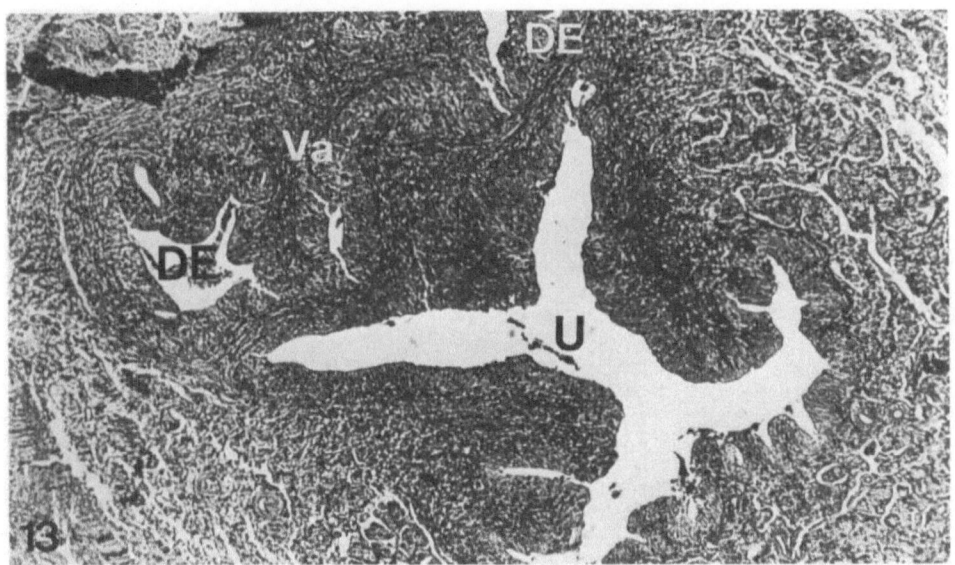

Fig. 13. Establishment of a vagina (*Va*) in a male adult animal the mother of which had been treated with cyproterone acetate from day 17 to day 20 of pregnancy. The vagina is lined by a mainly stratified epithelium. Some glandular structures are recognizable (*arrows*). *DE*, ejaculatory ducts; *U*, urethra. Azan blue, × 130

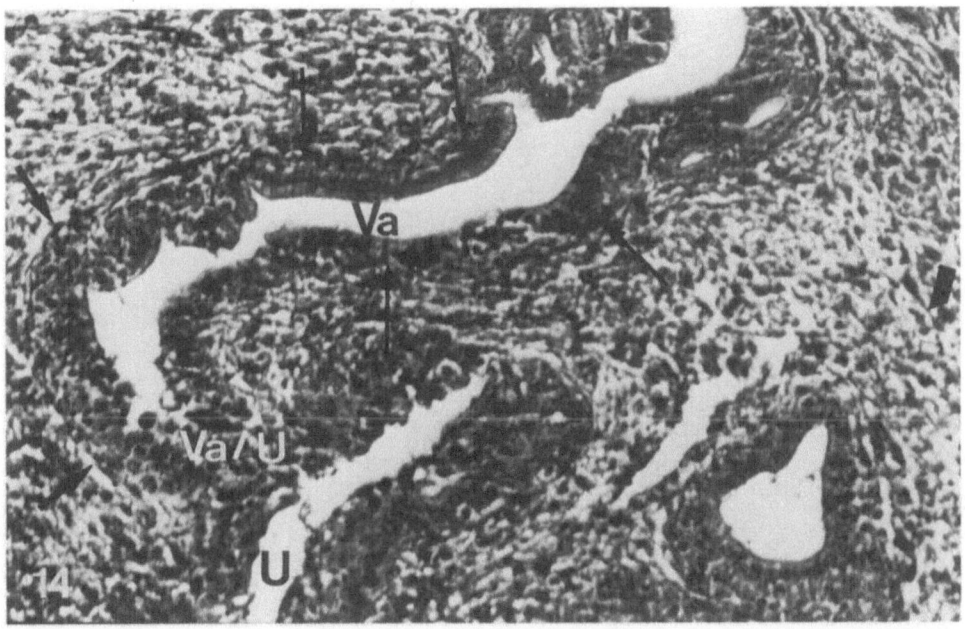

Fig. 14. Opening of the vagina (*Va*) into the urethra (*U*) of an adult male animal the mother of which had been treated with 10 mg cyproterone acetate from day 17 to day 20 of pregnancy. Areas with highly cylindrical epithelium (*small arrows*) are clearly distinguishable from areas with stratified epithelium (*large arrows*). Azan blue, × 130

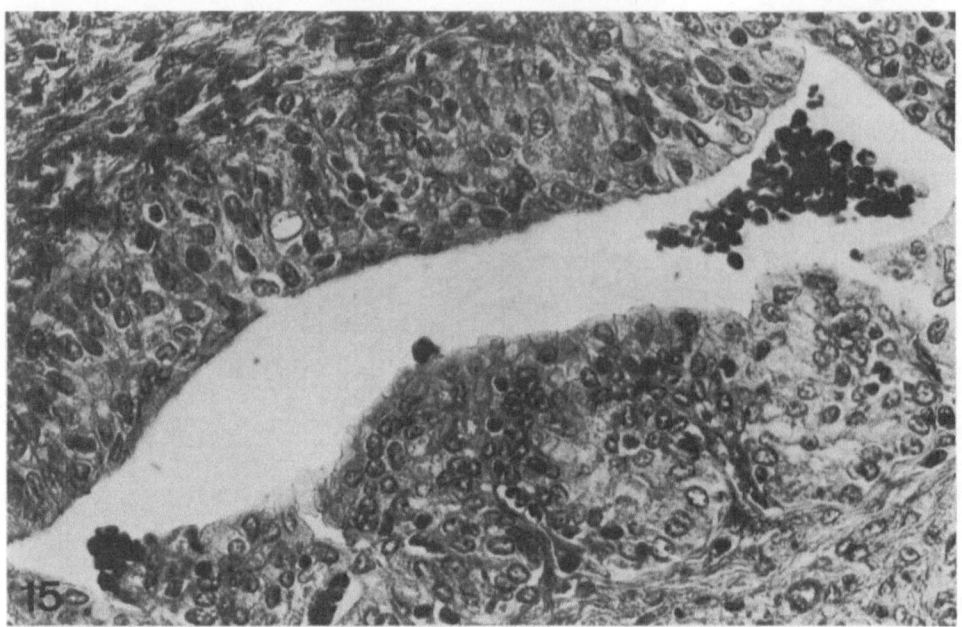

Fig. 15. Enlarged section of the vagina of an adult male animal the mother of which had been treated with 10 mg cyproterone acetate from day 17 to day 20 of pregnancy. Note the very marked development of the epithelium. Hematoxylin-eosin, × 640

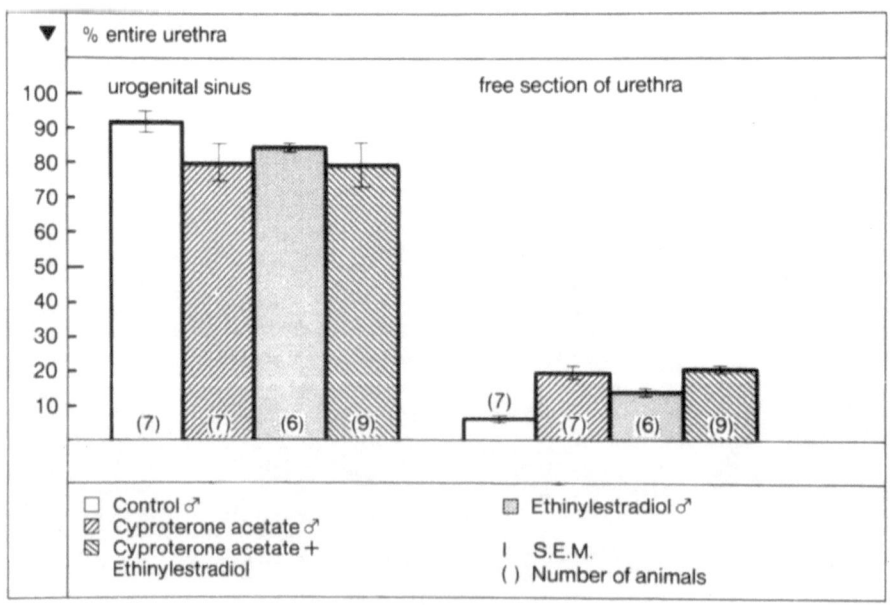

Fig. 16. The prenatal influence of cyproterone acetate, ethinyl estradiol, and ethinyl estradiol + cyproterone acetate on the length of the urogenital sinus and the free section of urethra as a proportion of the entire urethra of adult male rats

24

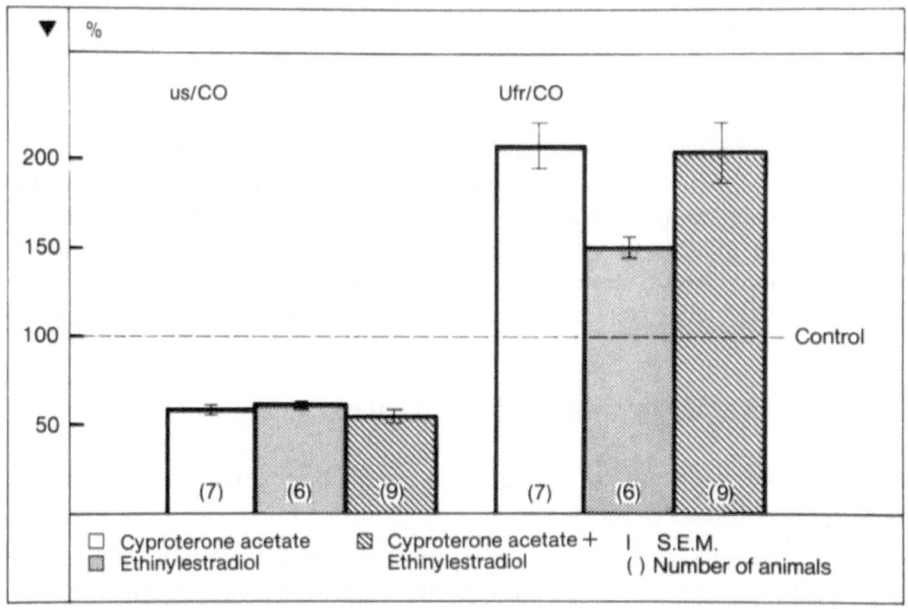

Fig. 17. The prenatal influence of cyproterone acetate, ethinyl estradiol, and ethinyl estradiol + cyproterone acetate on the length of the urogenital sinus (*US*) and the length of the free section of the urethra (*Ufr*) in relation to the controls

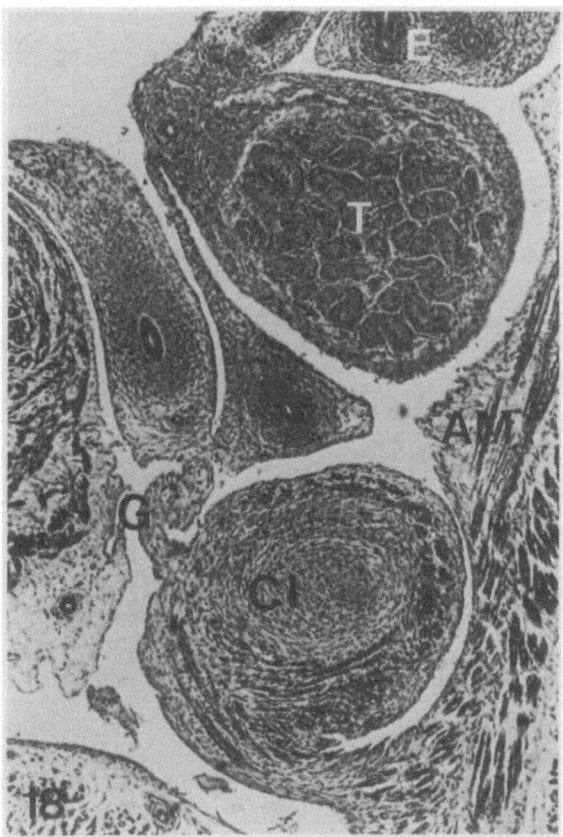

Fig. 18. Male fetus exposed to cyproterone acetate. Gubernaculum (*G*) and inguinal cone (*CI*) are normally developed. *AM*, abdominal muscles; *T*, testis; *E*, epididymis. Hematoxylin-eosin, × 64

25

### 3.3.1 Qualitative Macroscopic Examination of the Adult Animals

#### 3.3.1.1 Development of the External Genitals

*Days 1–13.* Again, the animals cannot be sexed at the time of birth because of the retarded development of the external genitals. Hypertrophied nipples are present in all animals at the time of birth. Nipples are always clearly recognizable, even in those animals having died in utero. The first differences between the sexes appear after 9 days: the phallus is further developed in the male animals than the females. The anlage of nipples is still very much pronounced in both sexes. Reliable sexing is possible only after 14 days.

Further Development of the External Genitals of the Male Animals

*Day 14.* The penis is only slightly developed in comparison to the control animals. The scrotal anlage is hardly recognizable.

*Day 16.* The nipples are no longer pronounced as in the female animals. The picture is otherwise unchanged and remains so until the 4th week of life, except that the nipples continue to regress.

*Days 20–24.* The scrotum in some animals continues to develop slightly during this period. The testes of all animals are still in the inguinal region.

*Day 27.* None of the animals display bilateral testicular descent by day 27: they display either unilateral descent into a extremely poorly developed scrotum or no descent at all. No scrotal anlage is present in these latter animals.

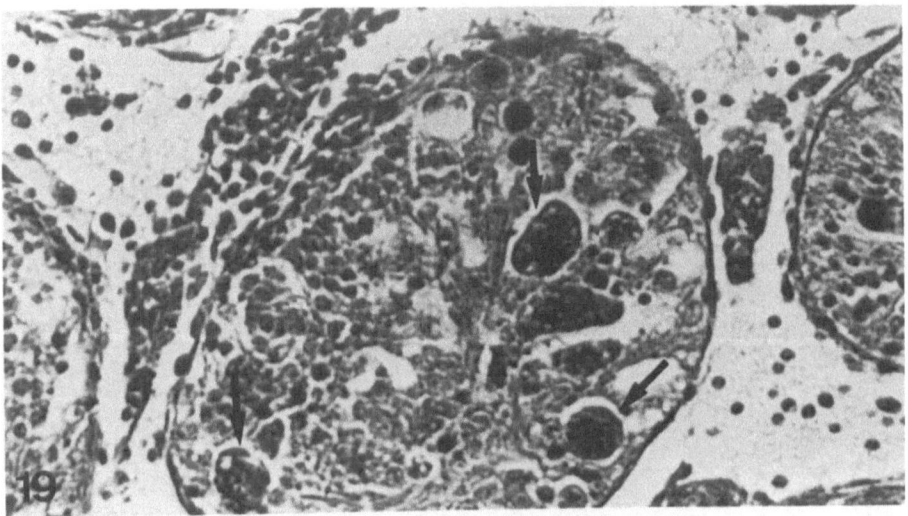

Fig. 19. Descended testis of an adult animal the mother of which had been treated with 0.03 mg ethinyl estradiol from day 17 to day 20 of pregnancy. The seminiferous tubules are almost completely degenerated. Within the lumen of the tubules multinucleated giant cells are visible (*arrows*). Also, a marked increase in the Leydig cell content is present. Hematoxylin-eosin, × 320

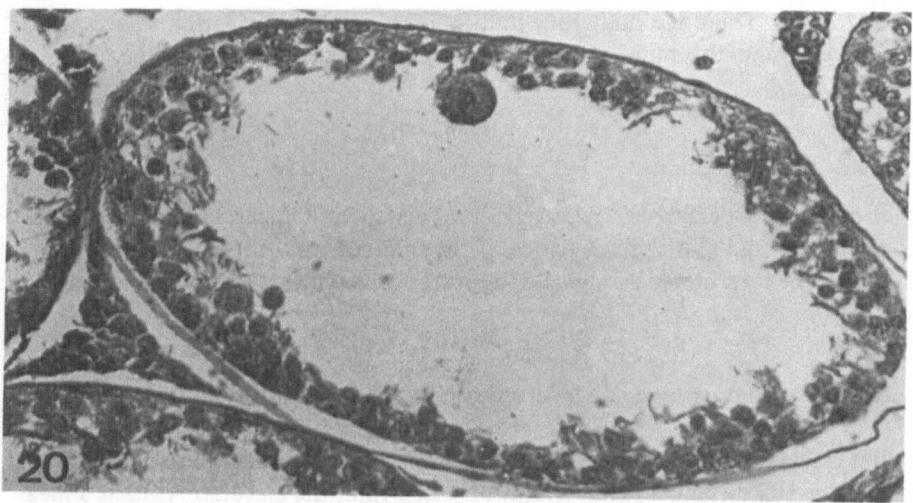

Fig. 20. Undescended testis of an adult animal the mother of which had been treated with 0.03 mg ethinyl estradiol from day 17 to day 20 of pregnancy. The seminiferous tubules are mainly bare of germinal epithelium. Only some spermatogonia and a few spermatocytes are recognizable. Again, multinucleate giant cells are developed. Hematoxylin-eosin, × 400

*Day 28.* Some of the animals display bilateral testicular descent into a small scrotum, but descent is unilateral in the majority of animals at this time.

*Days 29–33.* In the course of the next 5 days, the testes of six animals descend into a small scrotum, while unilateral descent occurs in another three (right side in one animal, left side in two). No scrotal anlage is present on the side on which descent has failed to take place. This picture does not alter up to the day of death (day 56 postnatally).

### 3.3.1.2 The Internal Genitals

Overall, the internal genitals of all six male animals studied display serious disturbances such as are well known to be induced by estrogens at high concentrations (Greene et al. 1939; Raynaud 1940, 1942; McLachlan and Newbold 1975; Nomura and Masuda 1980). The incompletely descended testes of two of the three animals with unilateral cryptorchidism are situated in a minute everted inguinal cone in the inguinal ring, while the undescended testis of the third animal lies directly in the abdominal cavity. Regardless of whether or not they have descended, all testes are smaller than normal and sometimes even so minute that they can only be regarded as rudiments.

### 3.3.2 Qualitative Microscopic Examination of the Adult Animals

Testes

The testes reveal disturbances of spermatogenesis, the extent differing according to the animal investigated as well as the section of the tubule. The undescended gonads are always completely degenerated. But in the descended gonad, too, the seminiferous tubules are often bare of germinal epithelium, and there are some isolated degenerating spermatocytes. There are numerous multinucleate giant cells within the lumen of the Sertoli cells, and the number of interstitial cells seems to be increased (Figs. 19 and 20).

Persistence of Müllerian ducts

Müllerian derivatives persit over relatively long distances. On the basis of the course of the derivatives and the characteristics of the epithelium with which the ducts are lined, they can be interpreted as vagina with posterior rudiments of a cervix (Figs. 21–24).

### 3.3.3 Quantitative Microscopic and Macroscopic Examinations of the Adult Animals

The genital tract of the male animals of this group achieves a mean length of 69.15 mm. The vasa deferentia display a mean length of 44.53 mm, the ejaculatory ducts a mean length of 1.57 mm, and the urogenital sinus a mean length of 23.02 mm. The free section of the urethra achieves a mean length of 4.01 mm (Table 3, Fig. 5).

The proportions of the individual sections of the genital tract are as follows: the vasa deferentia account for 64.40%, the ejaculatory ducts for 2.27%, and the urogenital sinus for 33.29% (Fig. 6, Table 4). In this group, the free section of the urethra and the urogenital sinus account for 14.84% and 85.16% respectively of the entire length of the urethra (Fig. 16).

A comparison of the absolute values of the individual sections of the genital tract with those of the control animals produces the following relationships. The length of the entire genital tract corresponds to only 79% of the control value. The

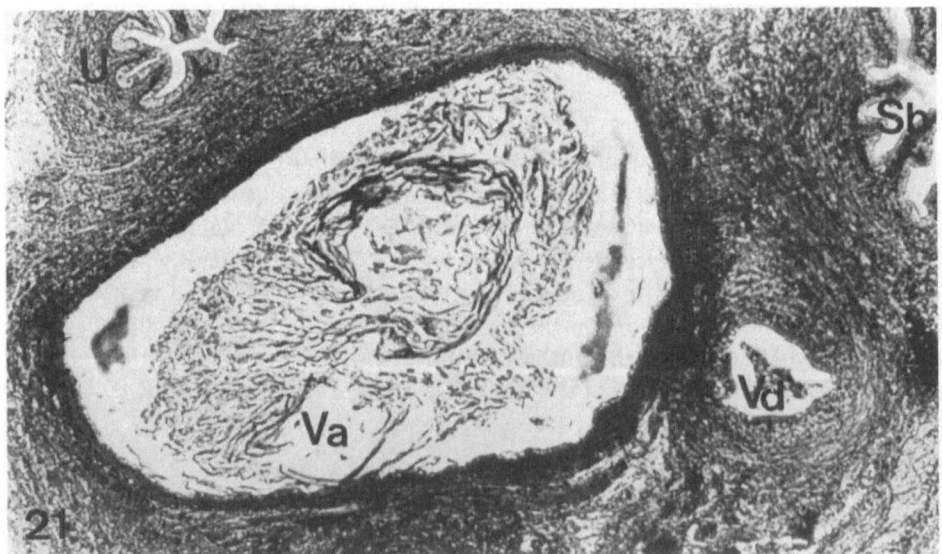

Fig. 21. Persistence of Müllerian derivatives. Development of a vagina (*Va*) of enormous size in an adult male animal the mother of which had been treated with 0.03 mg ethinyl estradiol. The epithelium is highly keratinized. *Sb*, ampulla of the seminal vesicle; *U*, urethra; *Vd*, vas deferens. Azan blue, × 50

28

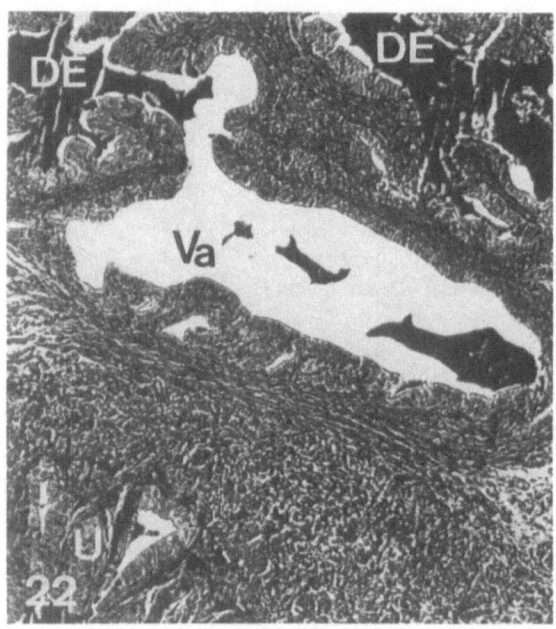

Fig. 22. Persistence of Müllerian derivatives. Opening of the right ejaculatory duct into the vagina (*Va*) of an adult male animal the mother of which had been treated with 0.03 mg ethinyl estradiol. *DE*, ejaculatory ducts; *U*, urethra. Azan blue, × 130

reduction in the length of the vasa deferentia is not significant, whereas that of the urogenital sinus is significant (64.90% of the normal value). The length of the ejaculatory ducts, on the other hand, increases considerably, being on average five times greater than that in the controls. At 151.89% of the control value, the proportion of the free urethra has also increased controls. At 151.89% of the control value, the proportion of the free urethra has also increased considerably (Table 3, Fig. 17).

### 3.3.4 Male Fetuses

The length of the genital folds in the fetuses of this group is only slightly increased in comparison to the male fetuses of group IIa (cyproterone acetate), while the length of the genital cord is slightly decreased. Overall, the mean length of the genital tract is not significantly different from that of group IIa, although the proportions have altered in comparison to the male fetuses of groups Ia and IIa − particularly within the genital cord. At 544 μm, in comparison to 462 μm in the controls, the length of the Wolffian ducts measured from the cranial pole of the genital cord up to the formation of the ejaculatory ducts has also increased in this group, although not by as much as in group IIa. The length of the ejaculatory ducts, on the other hand, has increased considerably (to 317 μm) in comparison both to the controls and to the fetuses of group IIa (Table 5). There is a distinct change in the distance by which the testes have descended. On average, they are situated only 1.289 μm from the caudal renal pole. None of the animals in this group display normal descent. In two animals the testes are still in the region of the caudal renal pole; in the other five they are either situated with their caudal pole at the level of the cranial neck of the bladder at the lateral abdominal wall, or they are still above the cranial pole of the bladder. In keeping with this, the gubernacula also display only slight development. Differentiation of the inguinal cone is largely normal (Table 5, Figs. 8 and 25).

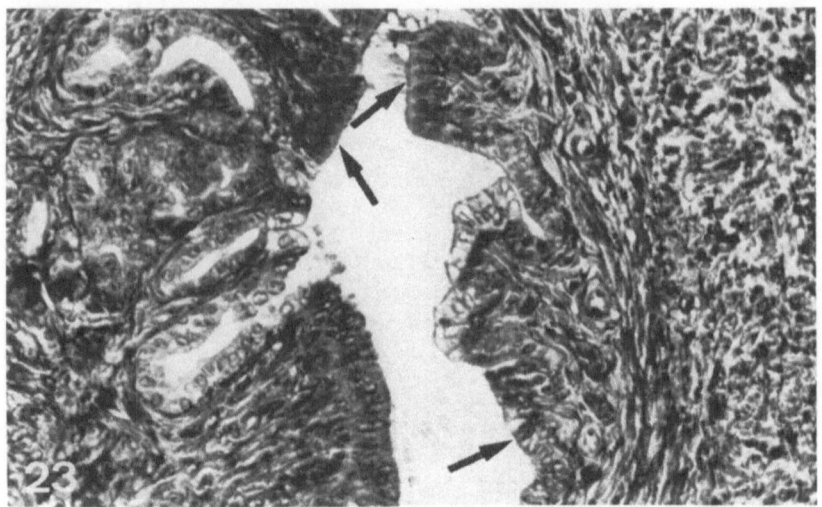

Fig. 23. Persistence of Müllerian derivatives. Enlarged section of the vagina (*Va*) of an adult male animal the mother of which had been treated with 0.03 mg ethinyl estradiol. Columnar epithelium typical of Müllerian derivatives is clearly recognizable (*arrows*). Furthermore, the glandular development is striking. Azan blue, × 320

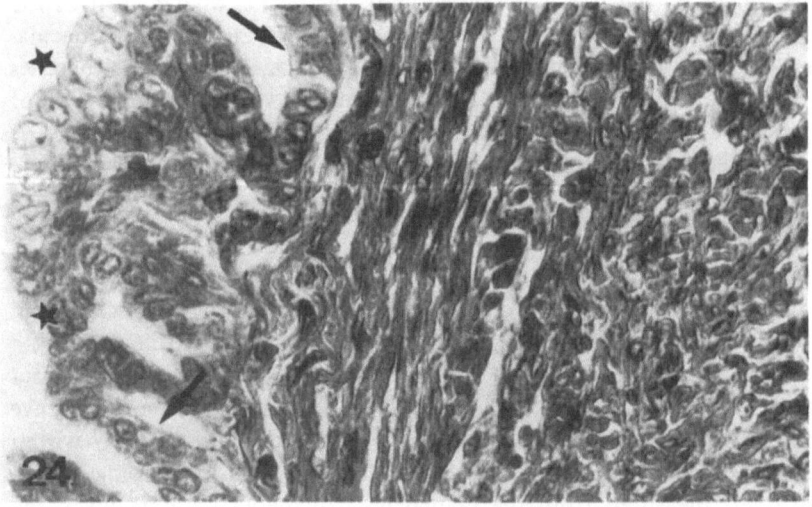

Fig. 24. Enlarged section of Fig. 23. Areas with columnar epithelium (*arrows*) and stratified epithelium consisting of two zones (*stars*) can be seen. Azan blue, × 640

## 3.4 Cyproterone Acetate and Ethinyl Estradiol

In the cyproterone acetate and ethinyl estradiol group (group IV), the development of the external genitals of the male animals from 15 litters was studied over a period of from 64 to 85 days. Qualitative macroscopic studies of the internal genitals were conducted in 17 adult male animals, while quantitative studies (microscopic and macroscopic) were conducted in nine male adult animals. Qualitative microscopic

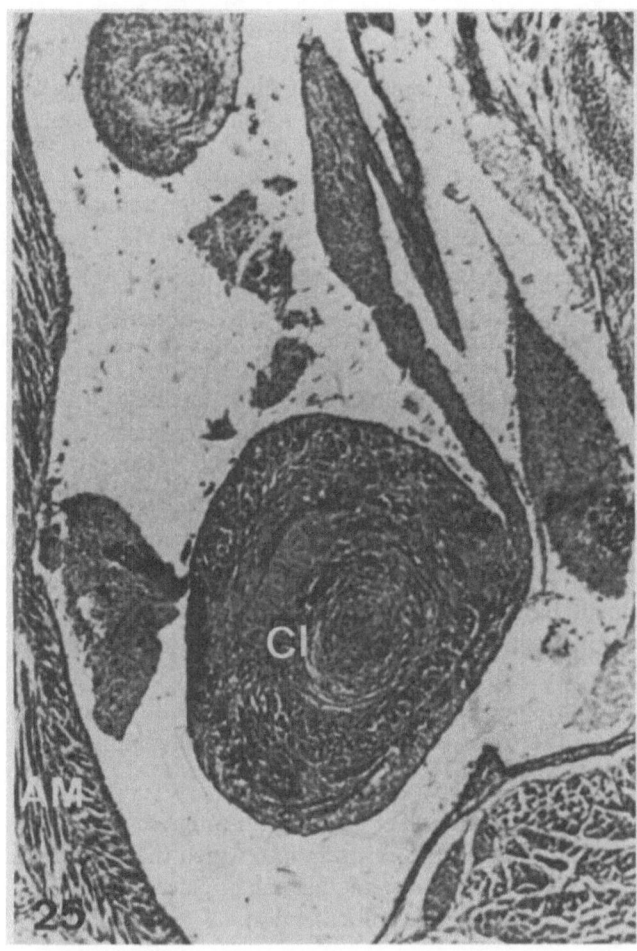

Fig. 25. Development of the gubernaculum (*G*) and the inguinal cone (*CI*) in a male fetus exposed to ethinyl estradiol. The gubernaculum in particular is only poorly developed. *AM*, abdominal muscles. Hematoxylin-eosin, × 64

examinations were performed in 15 male animals; eight male fetuses were examined, mainly from a quantitative point of view.

### 3.4.1 Qualitative Macroscopic Examination of the Adult Animals

#### 3.4.1.1 Development of the External Genitals

*Days 1–11.* Sexing is again impossible at first in this group. Externally the male and female animals are identical to the animals in group III. The hypertrophy of the nipples regresses in all animals during the next few postnatal days. Sexing is possible after 8 days in a few animals. In these cases, the phallus is again much further developed in the male animals than in the females.

Further Development of the External Genitals of the Male Animals

*Days 12–14.* After 14 days the nipples are still present (but only just discernible) in the inguinal region. There is still no sign of a scrotal anlage at this time. The penis is poorly developed.

*Day 15.* The first signs of a scrotal anlage are now recognizable in some animals. This picture changes only slightly during the next few days.

*Day 19.* The scrotum is still rudimentary even after 19 days, and then in only a few animals: the majority display no sign of a scrotum at all. The testes of some animals are in the inguinal region.

*Days 20–26.* The external picture of the genitals remains virtually unchanged during this period. Development of the scrotal anlage remains poor in most animals. In none of the animals have the testes descended by day 26.

*Day 27.* Development progresses somewhat in the course of the next 24 h. Of 18 animals 6 have developed a minute scrotum in this period, and the gonads of three animals are also within the scrotum – the left always somewhat higher than the right. The testes of two further animals have descended only on the left side, while no descent at all has taken place in one animal. In another seven animals the scrotum is recognizable only as a shallow protrusion. The testes are also within this region in some of these animals, while they are indiscernible in the others. The other five animals exhibit no scrotum at all, and the testes are visible in the inguinal region in only one of these animals.

*Days 28–33.* Both the degree of scrotal development and descent progress somewhat: after almost 5 weeks, seven animals display a small scrotum into which the testes have descended. Unilateral descent, always of the right testis, has taken place in eight animals; three still display neither descent nor any signs of a scrotum.

*Days 34–69.* After 39 days the testes of the majority of animals have descended into a sometimes very small scrotum. Only right-sided descend has taken place in others. One animal is devoid of a scrotum. The testes are easily visible in the inguinal region.

*Days 70–85.* This situation remains unchanged up to the day when the animals are killed. Of a total of 17 surviving male animals, ten display bilateral testicular descent into a small scrotum. The testes of six animals have descended unilaterally on the right side, and the scrotum is also present only on the right. One animal has developed no scrotum at all and its testes are in the inguinal region. All animals except two have developed nipples, some of which are very pronounced.

### 3.4.1.2 Internal Genitals

All animals display severe malformation of the entire genitals. In most cases the testes are very small. In those cases in which no or only unilateral descent has taken place the testes are situated – as in group III – in a minute peritoneal pouch.

### 3.4.2 Qualitative Microscopic Examination of the Adult Animals

Testes

The germinal epithelium of the testis is found to be severely damaged (Figs. 26—28). The descended testes of animals with unilateral cryptorchidism show a partial thickening of the peritubular connective tissue. The Sertoli cells are characterized by a marked content of lipoid droplets as well as many lysosomes. Furthermore, a lot of defective spermatids point to the disturbed function of the gonads (Figs. 29—31). There are some changes in the blood-testis barrier, like a conspicuous folding of the membrane system, but in fact the blood-testis barrier is functionally intact (Fig. 32). The Leydig cells appear to be normally differentiated.

In the undescended gonad the signs of degeneration are far more pronounced. There are a lot of degenerating germ cells within the Sertoli cell which have obviously been phagocytosed. The germ cells are mainly developed up to the stage of primary spermatocytes only. The peritubular wall is thickened and hyalinized due to an increased incorporation of collagen fibers and the deposits of amorphous substance (Figs. 33—35).

In spite of striking changes, in principle the blood-testis barrier is again functionally intact (Fig. 36).

Persistence of Müllerian ducts

As reported for group III, Müllerian ducts persist for relatively long distances. In principle, their course is identical to that of the persistent Müllerian derivatives described for group III (Figs. 37 and 38).

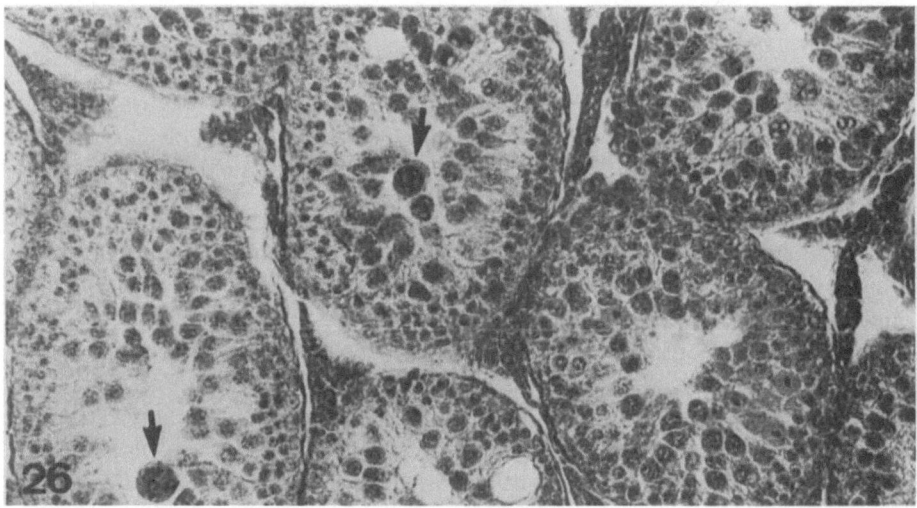

Fig. 26. Descended testis of an adult animal the mother of which had been treated with 10 mg cyproterone acetate + 0.03 mg ethinyl estradiol from day 17 to day 20 of pregnancy. Only spermatogonia and early spermatocytes are recognizable. Also, multinucleated giant cells (*arrows*) have been formed. The Leydig cells appear to be increased. Hematoxylin-eosin, × 320

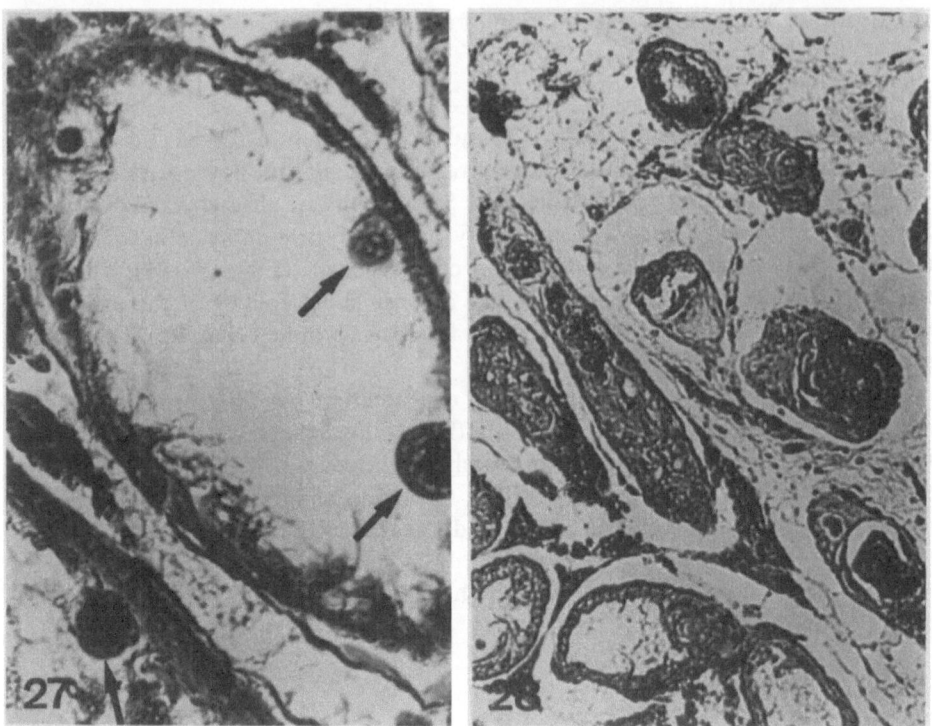

Fig. 27. Descended testis of an adult animal the mother of which had been treated with 10 mg cyproterone acetate + 0.03 mg ethinyl estradiol from day 17 to day 20 of pregnancy. The seminiferous tubules are almost completely depopulated of germ cells apart from multinucleated giant cells (*arrows*). Hematoxylin-eosin, × 320

Fig. 28. Complete necrosis of the undescended testis of an animal the mother of which had been treated with 10 mg cyproterone acetate + 0.03 mg ethinyl estradiol from day 17 to day 20 of pregnancy. Azan blue, × 320

### 3.4.3 Quantitative Microscopic and Macroscopic Examinations of the Adult Animals

At a length of 72.18 mm, the genital tract has achieved only 82% of the control value. Within the genital tract, the length of the vasa deferentia at 50.77 mm remains unchanged in comparison to the controls. In contrast, the urogenital sinus is considerably shorter — as in group II and group III, although it is even more pronounced than in these groups. At 20.02 mm it achieves only 56% of the control value. The ejaculatory ducts are much longer, achieving a mean length of 1.32 mm or almost four-and-a-half times that of the controls. The length of the free urethra is also considerably greater than in the controls — 5.42 mm compared to 2.64 mm (Table 3, Figs. 5 and 17). The proportions within the genital tract have likewise changed considerably. The proportions of the vasa deferentia and the ejaculatory ducts have increased, while that of the urogenital sinus has decreased, resulting in a ratio of 70.34% (vasa deferentia) to 1.83% (ejaculatory ducts) to 27.74% (urogenital sinus), compared to 60.04% (vasa deferentia), 0.34% (ejaculatory ducts), and 40.53% (urogenital sinus) in the control animals (Table 4, Fig. 6). The shifts within the urethra correspond to those of group III. The urogenital sinus decreases in length to 78.69% of the entire urethra and the free portion of the urethra increases to 21.31% (Fig. 16).

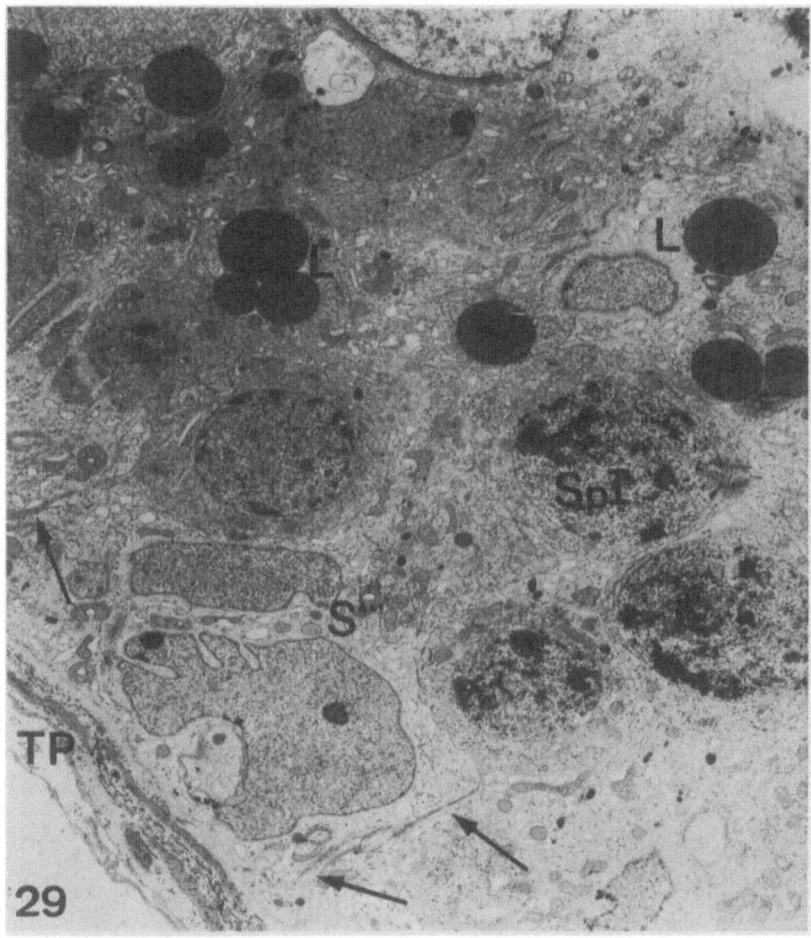

Fig. 29. Ultrastructure of a seminiferous tubule from the descended testis of an adult animal the mother of which had been treated with 10 mg cyproterone acetate + 0.03 mg ethinyl estradiol with unilateral cryptorchidism. The blood-testis barrier appears to be normally differentiated (*arrows*). There is a conspicuous broadening and collagenisation of the tunica propria (*TP*). *S*, Sertoli cell with typically invaginated nucleus; *Sp* I, spermatocyte I; *L*, lipoid droplets. × 3300

### 3.4.4 Male Fetuses

Inhibition of descent is even greater than in group IIIa. The cranial testicular pole lies an average of 902 $\mu$m beneath the caudal renal pole (Table 5), Fig. 8). Although the testes appear thin and elongated, no anomalies are detectable in the histological section. The gubernacula are poorly developed (Fig. 39). The length of the genital folds is the same as in the male animals of group IIIa, while that of the genital cord at 1138 $\mu$m has increased distinctly in comparison to group IIa and group IIIa. The length of the ejaculatory ducts has increased in the region of the genital cord, achieving a mean value of 418 $\mu$m compared to 317 $\mu$m in group IIIa, 223 $\mu$m in group IIa, and 182 $\mu$m in group Ia (Table 5, Fig. 5). The length of the Wolffian ducts measured from the anterior end of the genital cord up to the junction of the Wolffian ducts and seminal vesicles with the ejaculatory ducts is 719 $\mu$m, and is therefore likewise

35

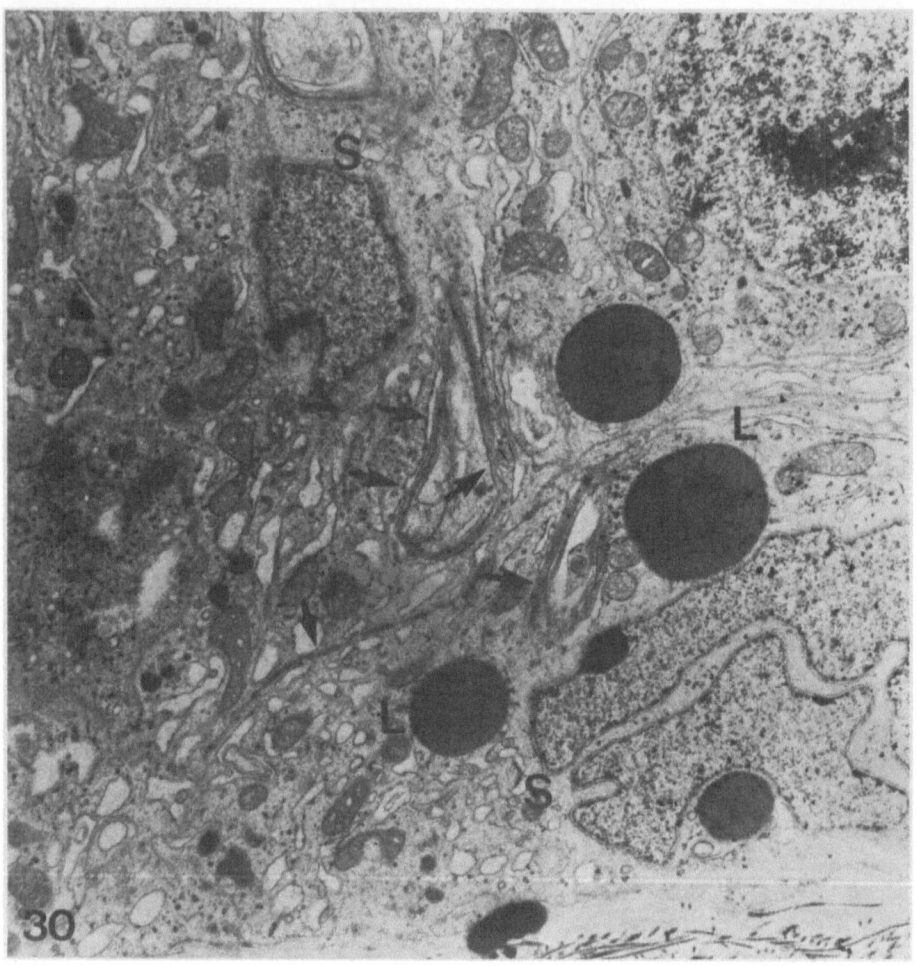

Fig. 30. Ultrastructure of a seminiferous tubule from the descended testis of an adult animal with unilateral cryptorchidism the mother of which had been treated with 10 mg cyproterone acetate + 0.03 mg ethinyl estradiol from day 17 to day 20 of pregnancy. The lipoid droplets (*L*) seem to be increased. The blood-testis barrier is normally differentiated (*arrows*). *S*, Sertoli cell. × 9600

distinctly greater than in the controls, in which a length of 462 $\mu$m was measured. The entire length of the genital tract at 2427 $\mu$m is likewise increased in comparison to both the control animals and groups IIa and IIIa (Table 5, Fig. 39).

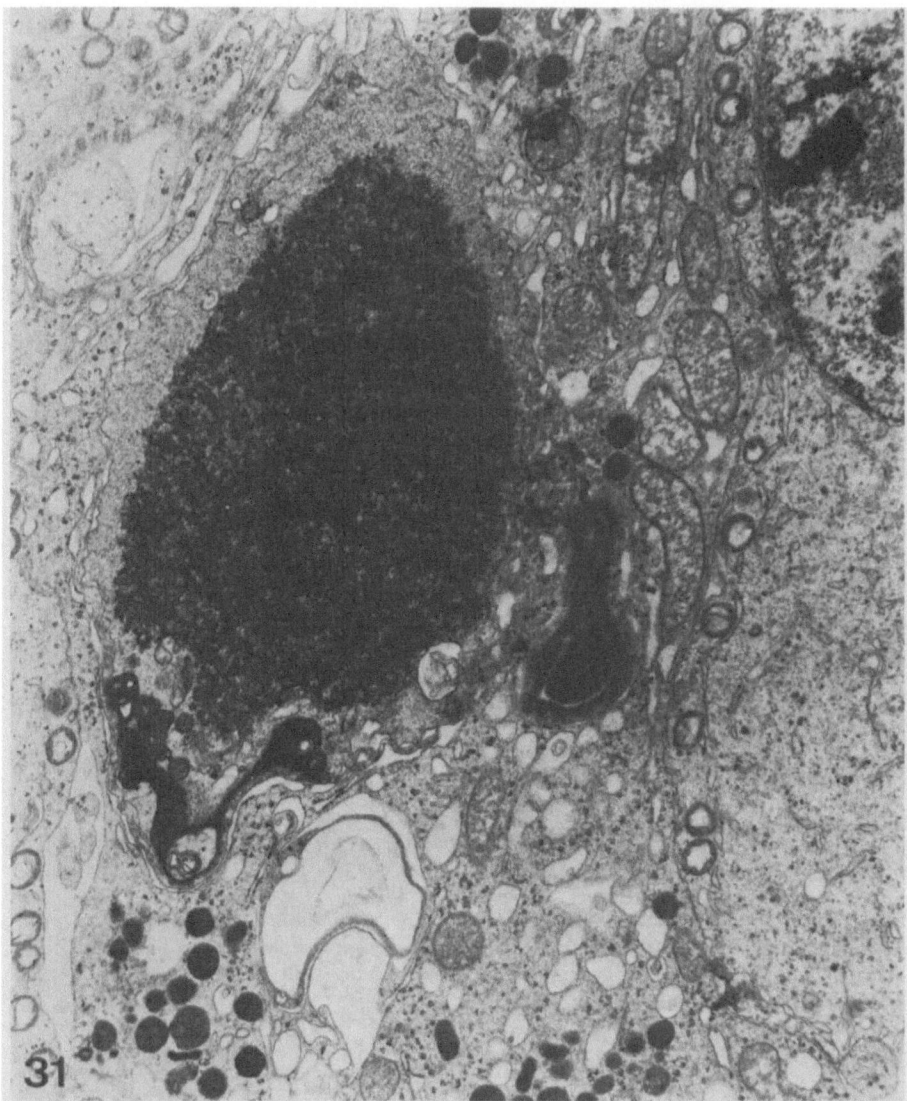

Fig. 31. Spermatid from the descended testis of an adult animal with unilateral cryptorchidism, the mother of which had been treated with 10 mg cyproterone acetate + 0.03 mg ethinyl estradiol from day 17 to day 20 of pregnancy. Both the development of the acrosome and the condensation of the nucleus are disturbed. × 18 000

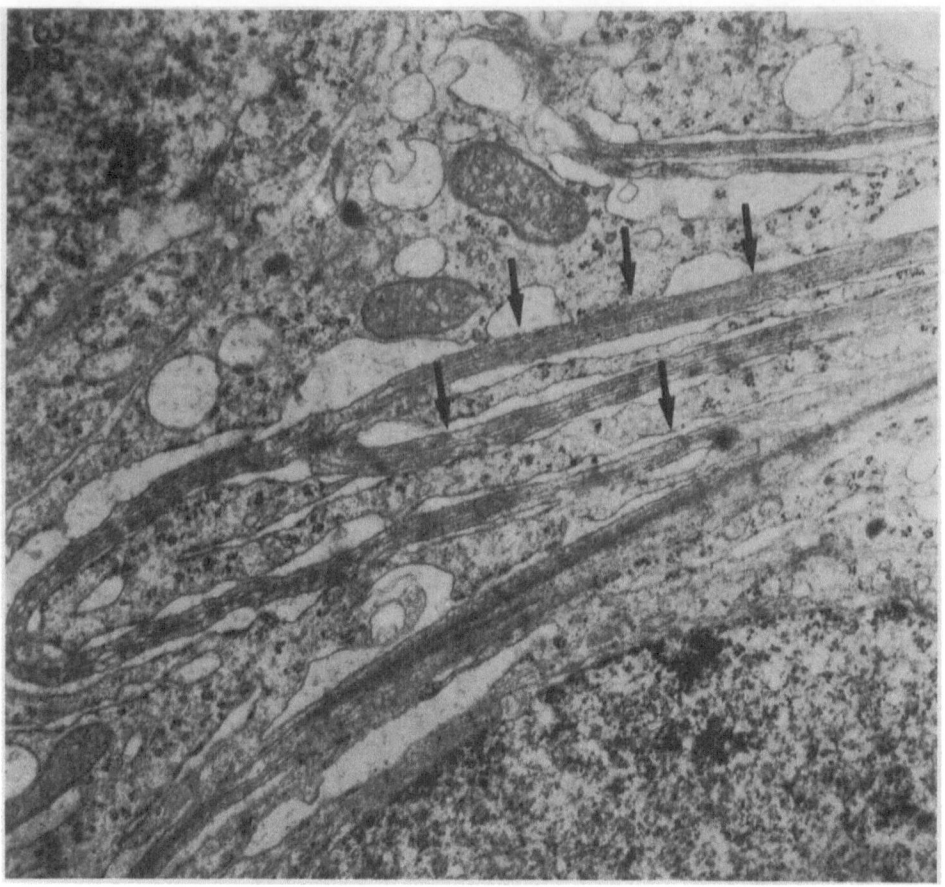

Fig. 32. Blood-testis barrier from the descended testis of an adult animal with unilateral cryptorchidism, the mother of which had been treated with 10 mg cyproterone acetate + 0.03 mg ethinyl estradiol from day 17 to day 20 of pregnancy. A slight dilation and folding of the blood-testis barrier (*arrows*) are recognizable. × 20 400

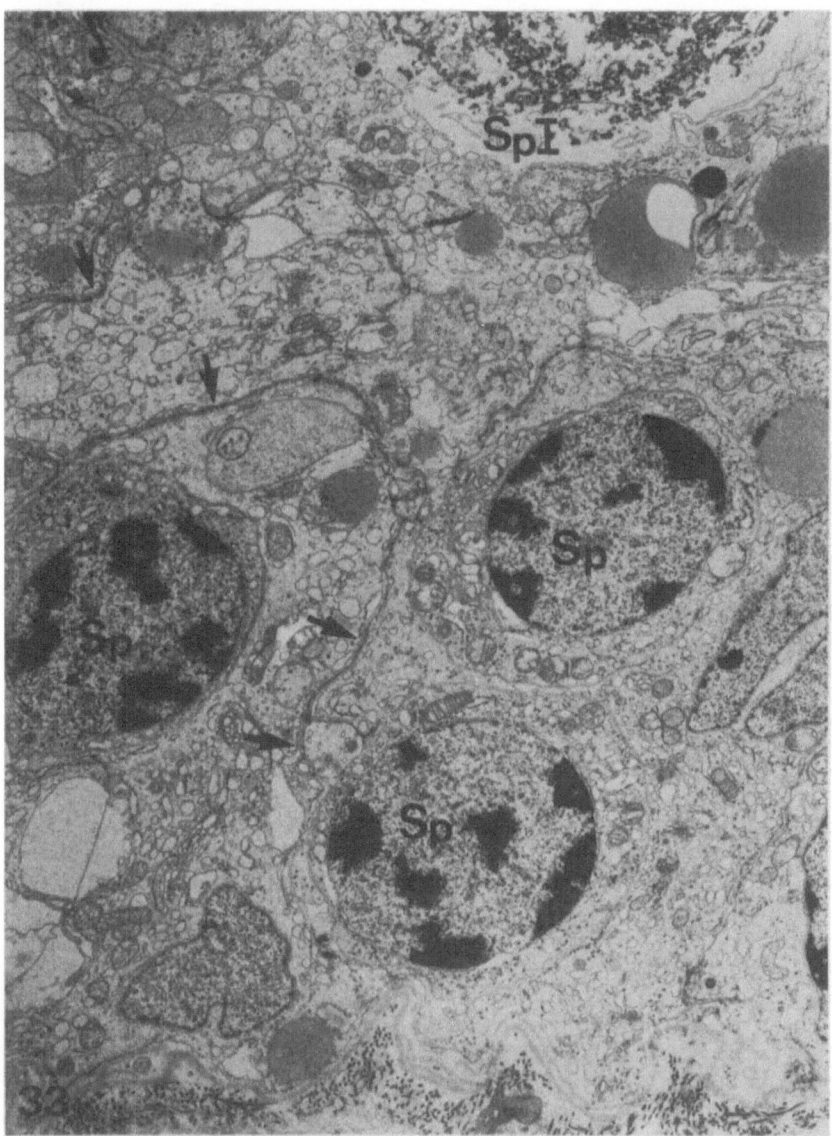

Fig. 33. Seminiferous tubule from the undescended testis of an adult animal with unilateral cryptorchidism, the mother of which had been treated with 10 mg cyproterone acetate + 0.03 mg ethinyl estradiol from day 17 to day 20 of pregnancy. The blood-testis barrier reveals no peculiarities (*arrows*). A degenerating spermatocyte is demonstrated. *Sp*, spermatogonia; *Sp I*, spermatocyte I. × 5700

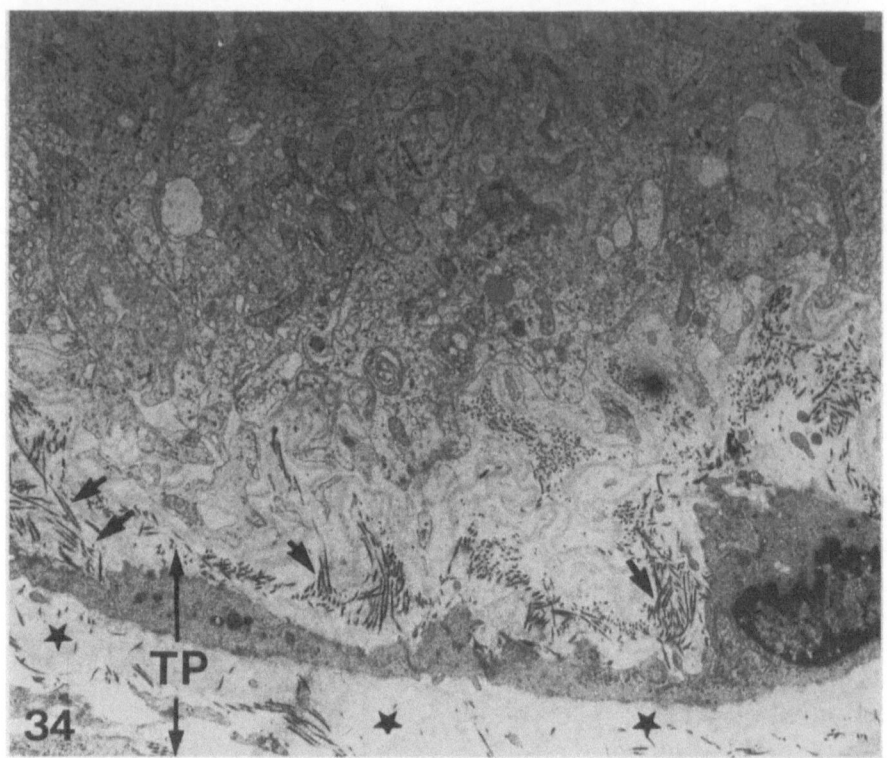

Fig. 34. Basal compartment of a seminiferous tubule from the undescended testis of an animal with unilateral cryptorchidism, the mother of which had been treated with 10 mg cyproterone acetate of 0.03 mg ethinyl estradiol from day 17 to day 20 of pregnancy. There is enlargement and hyalinization of the tunica propria (*TP*). Both an increased incorporation of collagen fibers (*arrows*) and amorphous material (*stars*) are to be noted. × 5700

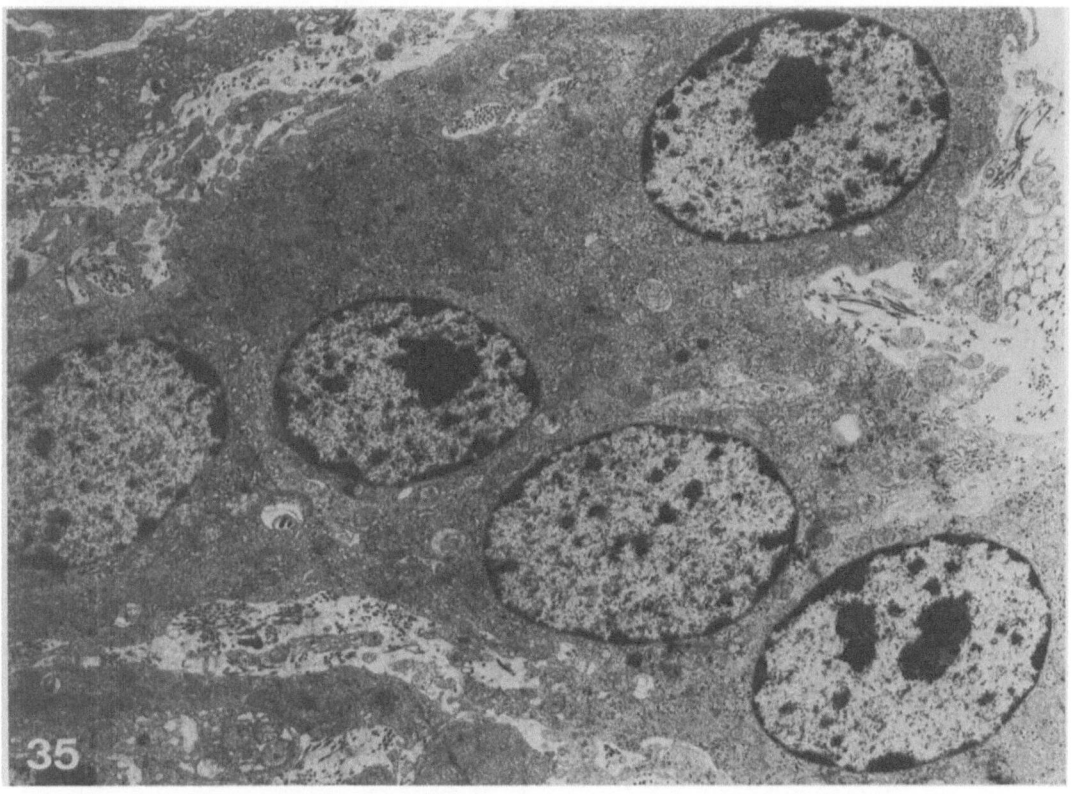

Fig. 35. Leydig cells in the undescended testis of an animal with unilateral cryptorchidism, the mother of which had been treated with 10 mg cyproterone acetate and 0.03 mg ethinyl estradiol from day 17 to day 20 of pregnancy. The nuclei show the normal peripheral arrangement of the heterochromatin. The surface appears to be poorly differentiated — a sign of low activity. × 6300

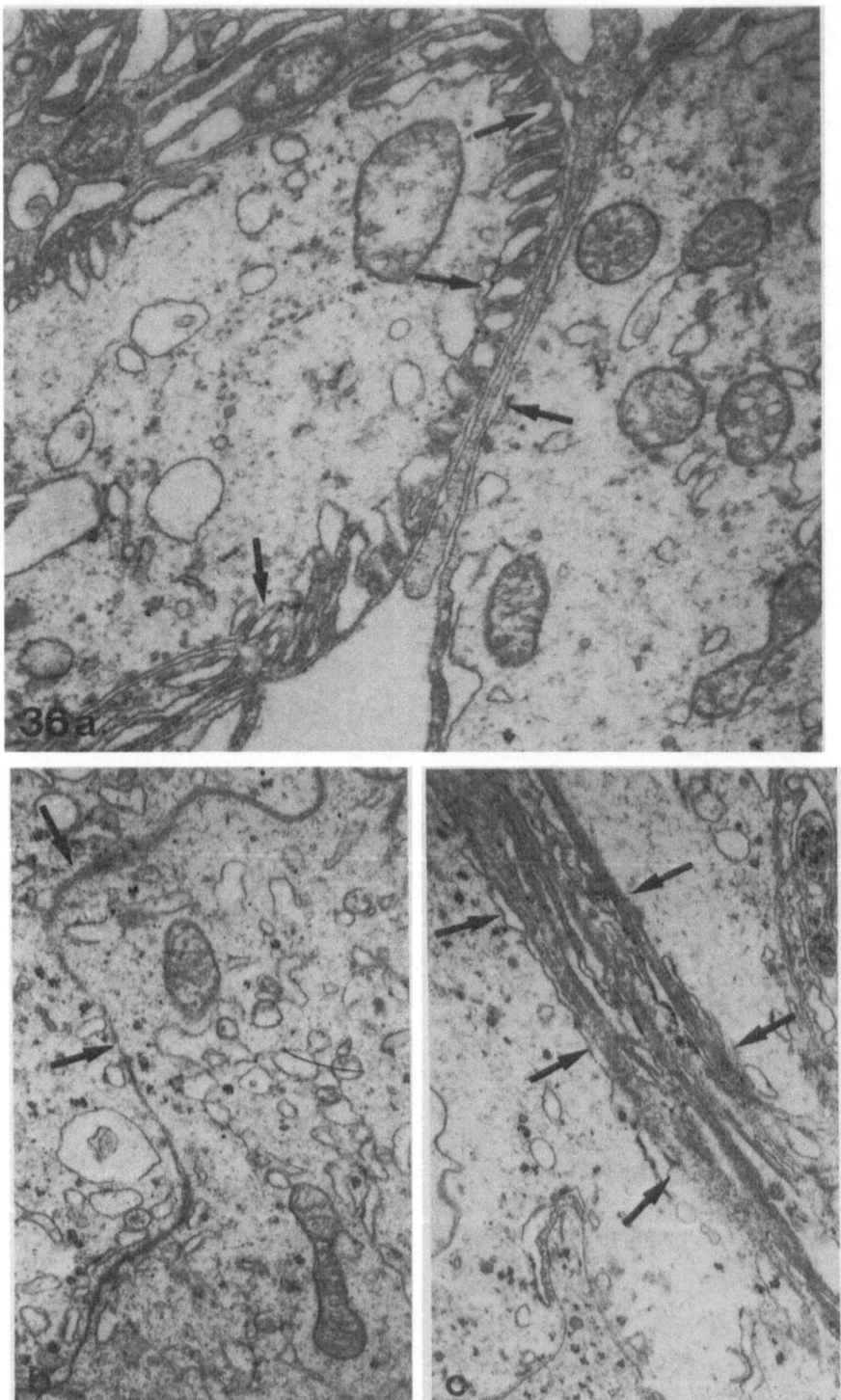

Fig. 36a–c. Blood-testis barriers from the undescended testes of animals with unilateral cryptorchidism, the mothers of which had been treated with 10 mg cyproterone acetate and 0.03 mg ethinyl estradiol from day 17 to day 20 of pregnancy. *a* conspicuous folding and partial fragmentation of the blood-testis barrier (*arrows*), × 22 800; *b* incomplete development of the blood-testis barrier (*arrows*), × 20 600; *c* enlargement of the blood-testis barrier (*arrows*), × 21 400

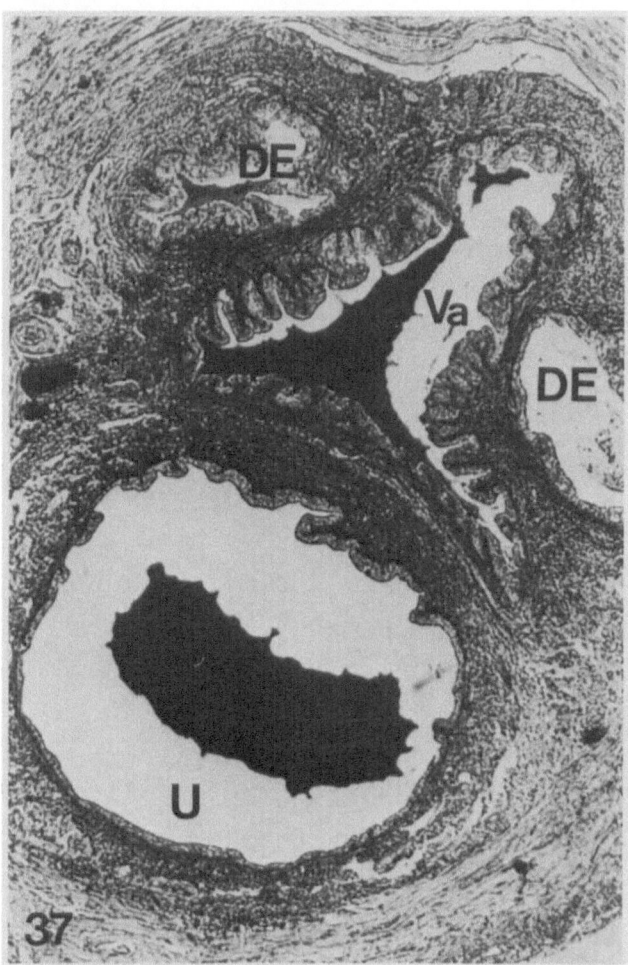

Fig. 37. Persistence of Müllerian derivatives. Development of a vagina (*Va*) in a male animal the mother of which had been treated with 10 mg cyproterone acetate + 0.03 mg ethinyl estradiol. *DE*, ejaculatory duct; *U*, urethra. Azan blue, × 64

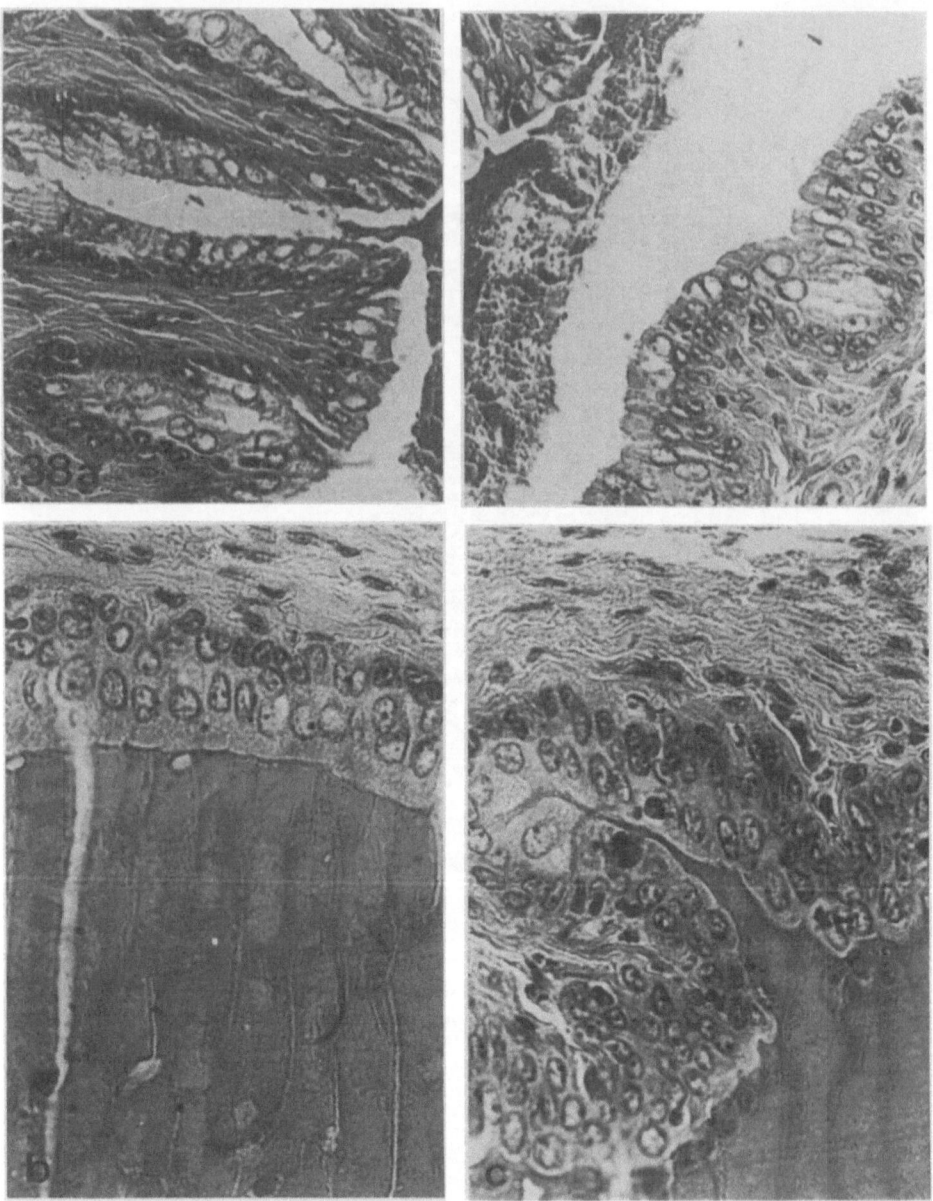

Fig. 38a–c. Characterization of the epithelium of the persistent vagina in male animals the mothers of which had been treated with 10 mg cyproterone acetate + 0.03 mg ethinyl estradiol from day 17 to day 20 of pregnancy. *a* enlarged section of the vagina revealing a two-layer epithelium and signs of glandular development; *b* enlarged section of the two-layer vaginal epithelium typical of the transient stage during the development of the vagina; *c* enlarged section of the vagina lined with stratified epithelium. Hematoxylin-eosin, × 640

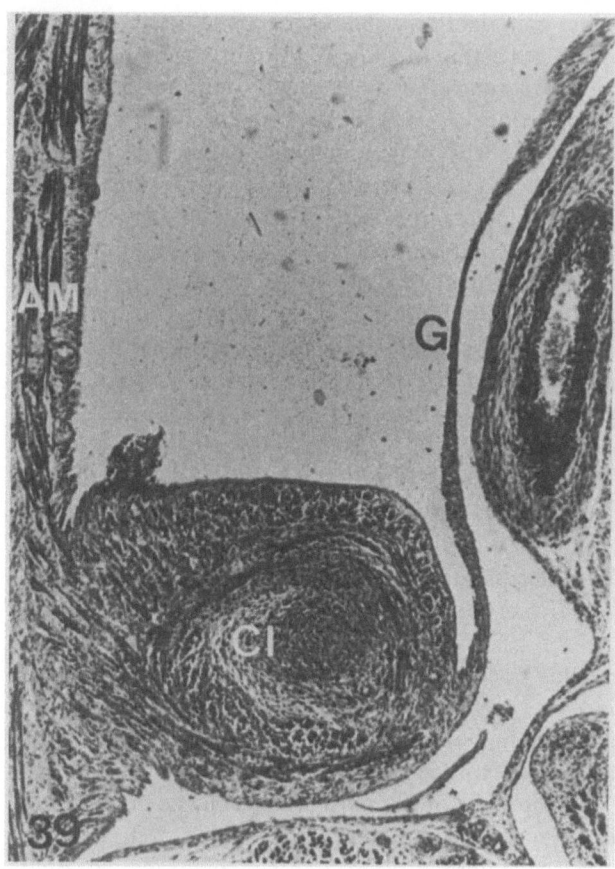

Fig. 39. Development of the gubernaculum (*G*) and inguinal cone (*CI*) in a male fetus exposed to cyproterone acetate + ethinyl estradiol. The gubernaculum in particular is only poorly developed. *AM*, abdominal muscles. Hematoxylin-eosin, × 64

# 4 Discussion

The object of the present paper was to obtain further information about a possible association between altered proportions within the genital tract of male rats on the one hand and the processes of testicular descent, cryptorchidism, or incomplete descent on the other with the aid of specific hormone treatments in gestating rats during the phase of fetal sexual differentiation.

In comparison to the controls, there are no significant quantitative differences between the treated groups as regards the effects of ethinyl estradiol, cyproterone acetate, and the combination of the two hormones on the development of androgen-dependent structures. The length of the Wolffian ducts is not significantly different from that of the controls. The reduction in the length of the genital tract is primarily a result of a reduction in the length of the urogenital sinus. Cyproterone acetate and ethinyl estradiol have an equal effect, which is neither reduced nor increased in combination. Although the length of the ejaculatory ducts is considerably greater than that of the controls, it does not differ significantly between the three treated groups. The decrease in the length of the urogenital sinus is accompanied by an increase in the length of the free section of the urethra, which is most pronounced under the influence of cyproterone acetate; the combination has no synergistic effect, and the increase in length brought about by ethinyl estradiol alone is not quite as great.

Changes in the individual sections of the genital tract are apparent not only from an absolute point of view: their proportions — particularly those of the ejaculatory ducts and the urogenital sinus — also change within the individual groups. The absolute reduction of the urogenital sinus is reflected in a reduction of the percentage of the genitals accounted for by the urogenital sinus, which is relatively similar for all three groups, although a slight synergistic effect is apparent under the combination. The situation is somewhat different as regards the ejaculatory ducts. Their proportion of the genitals is only slightly increased under the influence of cyproterone acetate, but by no means as drastically as under the effect of ethinyl estradiol and the combination of ethinyl estradiol and cyproterone acetate. The proportion of the Wolffian ducts does not increase to any particular extent.

Very distinct group differences also appear in a correlation between the entire urethra, the free section of the urethra, and the urogenital sinus. In this case, the increase in the proportion of the free urethra and the corresponding reduction in that of the urogenital sinus are most pronounced under the influence of cyproterone acetate. This effect remains unchanged in the combination, and is not so pronounced under ethinyl estradiol. The alterations in the proportions of the genitals discussed above refer to the adult male animals.

The situation obtaining at the time of birth cannot simply be extrapolated to the situation in the adult animals, since — even in a normal case — the proportions of the genitals at the time of birth alter in the course of sexual maturation. Quite apart from this, the fetuses had been exposed to a considerably higher dose of ethinyl estradiol than the animals which were brought on. It must nevertheless be noted that a distinct increase in length of the ejaculatory ducts is demonstrable even at birth. Under the influence of the higher estrogen dose, differences are recognizable not only in comparison to the controls, but also between the groups: the combination elicits the greatest increase in the length of the ejaculatory ducts, cyproterone acetate the least.

The combination sometimes elicits synergistic effects in the case of the other sections of the fetal genital system as well. The length of the genital cord, in particular, is increased to an ever greater extent under the combination of ethinyl estradiol and cyproterone acetate than under individual treatment. The length of the genital tract as a whole, which is already considerably increased under individual treatment in comparison to the controls, increases even more under the combination. However, neither the genital cord nor the genital tract as a whole achieve the length of the same structures in the female control animals.

Thus, taken overall, a shift of the junctions of the individual genital sections has taken place in a caudal direction, the proportional shifts within the genitals varying in the different groups. Of the sections most subject to changes from a quantitative point of view (ejaculatory ducts, urogenital sinus, free urethra), ethinyl estradiol has the greatest effect on the ejaculatory ducts, cyproterone acetate the greatest effect on the free urethra, while both hormones have the same effect on the urogenital sinus. Despite this, however, no correlation can be established between the altered proportions of the genitals and the degree of disturbance of testicular descent.

Further information about testicular descent may, however, be forthcoming from morphological evaluation of the prenatal and postnatal course of this process following prenatal treatment of the maternal animals with various hormones. A prerequisite for undisturbed testicular descent is a normal development of the ligamentous apparatus (gubernaculum, inguinal cone). Androgens are apparently not responsible for the development of these structures, since disturbed development of the ligamentous apparatus cannot be achieved with antiandrogens. Estrogens, on the other hand, cause inhibition of the gubernacula and inguinal cone. These findings agree with those of other study groups (Richter 1973; Neumann et al. 1970). Since estrogens can also suppress regression of the Müllerian ducts at the same time, it has been suggested — as mentioned in the Introduction — that Factor X (AMH) is also responsible for descent, but that androgens are not.

Our own results have confirmed this relationship, too. Ethinyl estradiol seems to be able to inhibit Factor X to a remarkable degree, like estradiol-17 $\beta$. The stronger inhibition of the regression of the Müllerian ducts is accompanied by a simultaneous disturbance of testicular descent.

In the fetuses the combination of cyproterone acetate and ethinyl estradiol nevertheless displays a distinct synergistic effect in comparison to the individual treatment with regard to disturbances of testicular descent. In contradiction of other results, testicular descent does not remain uneffected even by treatment with cyproterone acetate alone. Even considering that the individual values for the degree of descent vary greatly and that the internal proportions alter as a result of the somewhat lesser growth of the fetuses, it must nevertheless be stated that the position of the testes is not usually normal, despite normal development of the gubernaculum. They are situated somewhat too high and are displaced laterally, i.e., they are not close up against the neck of the bladder at birth. Postnatally, further progress is tardy in all three treated groups, and comes to a normal conclusion only in group II. The other two treated groups display disturbances. In principle, the effect of ethinyl estradiol alone and the combination of ethinyl estradiol and cyproterone acetate can be regarded as identical in this respect. The testes of these two groups descended even further after discontinuation of the treatment — and it must be remembered that the fetuses had been exposed to a considerably higher dose of ethinyl estradiol than the animals which were brought on.

Since, however, at least a positional shift of the testes is present at the time of birth under the influence of cyproterone acetate, and postnatal descent is concluded at a much later date than in the controls, the question arises to what extent androgens are infact involved in this process.

In rats, marked biosynthesis of testosterone is resumed in the 2 postnatal week (Döhler and Wuttke 1975; Miyachi et al. 1973). This correlates very well with the postnatal course of descent: the scrotal anlage becomes visible in the 2nd postnatal week and the testes of most animals are in the inguinal region by day 19, descending completely in the course of the next few days.

Perhaps it is not, after all, a simple question of androgens or Factor X. Descent is also a developmental process comprising different interdependent phases, which may in turn be regulated by different inducers. The hypothesis might therefore be postulated that Factor X is responsible for the stabilization and differentiation of the gubernacula and inguinal cone, while androgens are responsible for the final phase of descent, i.e., for eversion of the inguinal cone into the cremaster sac and for the inguinal passage of the testes into the scrotum. Applied to this study, this would mean that with the dose of estrogen used the development of the gubernacula and the inguinal cone is disturbed but not prevented. Under the delayed onset of postnatal androgen biosynthesis, a more or less rudimentary inguinal cone then everts into the cremaster sac, in which the testes are situated. The development of the scrotum itself (Neumann et al. 1970) does not take place in isolation from the development of the cremaster sac in the rat. The testes do not therefore descend into a previously developed scrotum — the formation of the scrotum and the entry of the testes take place synchronously. It is also possible that the everting inguinal cone induces the further development of the scrotal anlage: since the inguinal cone is small, the scrotum also remains small or is completely absent. It is also conceivable that, apart from resuming late in the treated groups, the postnatal biosynthesis of androgens also varies quantitatively. If the prenatal influence of ethinyl estradiol were to result in lower testosterone biosynthesis, the consequence would be a concentration high enough to allow a certain but very much smaller degree of scrotal development. Now, differentiation of the ligamentous apparatus as a morphological prerequisite for descent was not inhibited by cyproterone acetate, so the developmental deficit could still be made good under the postnatal biosynthesis of androgens, which was likewise resumed late.

It must now also be considered that the determination of the development of the scrotal anlage and of the conclusion of descent cannot be put on exactly the same footing as the marked postnatal resumption of androgen production. The results for descent in the fetuses under the influence of cyproterone acetate also support the hypothesis that the perinatal and neonatal secretion of testosterone creates the conditions for determination processes which become manifest later under the influence of the resumed production of androgens.

Apart from the pertinent studies with antiandrogens, the clinical picture of testicular feminization also suggests that the development of the gubernacula and inguinal cone is not androgen-dependent. The testes in these genotypically male but phenotypically female people have descended as far as allowed. In contrast, a normally developed scrotum is usually present in boys with cryptorchidism — yet another phenomenon which casts doubt on the primary androgen dependence of descent.

On the other hand, however, earlier experiments in which rats were treated neonatally with cyproterone acetate have produced indications of incomplete testicular descent (Neumann et al. 1967). Moreover, the further the testes have descended, the more the treatment of cryptorchidism with HCG or LH-RH during the 1st year of life is likely to succeed. Complete absence of prenatal descent precludes postnatal initiation (Lipshultz 1976).

A final point worthy of discussion is the possibility that, in addition to androgens, Factor X also plays a role in the final phase of descent. Studies in rats have shown that Factor X is demonstrable up to postnatal day 21, although no one has yet been able to explain the biological relevance of this (Donahoe et al. 1976, 1977; Josso and Tran 1979).

The sensitivity of Müllerian ducts to AMH applies for a relatively short time period, i.e., from 14 to 16 days post coitum in the rat and up to a crown-rump length of 30—32 mm in human fetuses. In in vitro experiments it has not been possible to get the Müllerian duct to regress as a result of AMH before the so-called critical period. Thereafter, male Müllerian ducts irreversible degenerate, whether in the absence or presence of AMH, whereas female Müllerian ducts are no longer responsive to AMH. Nevertheless, anti-Müllerian activity in the human testis has been shown in male fetuses 7—32 gestational weeks old, as well as in newborns 27—40 gestational weeks old (Hosso 1971; Josso et al. 1977).

According to Josso (Josso et al. 1977), AMH is a fetal hormone whose synthesis begins at the time or immediately after differentiation of seminiferous tubules, before the appearance of Leydig cells. There appears to be no correlation between the prolonged anti-Müllerian activity of the fetal testes and the short period during which the Müllerian duct is responsive to AMH.

Moreover, Hadziselimovic (1977) mentions that the number of fetal Sertoli cells is considerably reduced and sometimes even nil immediately after birth in babies with cryptorchidism. Fetal Sertoli cells are usually demonstrable up to the 3rd postnatal month. Hadziselimovic himself does not discuss this finding at all, but deals exclusively with degenerative changes of the Leydig cells.

In my opinion, however, this finding offers highly significant support for the hypothesis that – in addition to a possible role of androgens – Factor X, the site of formation of which is believed to be precisely the Sertoli cells (Blanchard and Josso 1974), is of major importance for the normal process of testicular descent.

To summarize, there are some indications that Factor X is primarily responsible for the differentiation of the ligamentous apparatus and thus for the initial displacement of the gonads, while androgens are required either alone or in addition for the final phase of descent. This latter point is supported by the previously mentioned fact that the further the testes have descended, the more likely the treatment of cryptorchidism with HCG or LH-RH is to succeed.

The question of the regulatory mechanisms of testicular descent cannot be answered simply by establishing or ruling out androgen or Factor X dependence. There rather exists increasing evidence that we are confronted here with an extremely complicated developmental process, in the regulation of which several systems are involved.


# 5 Summary

Gestating rats were treated from day 17 to day 20 post coitum with ethinyl estradiol, cyproterone acetate, and a combination of cyproterone acetate and ethinyl estradiol, after which the male young were studied, partly as fetuses and partly as adult animals.

The object of the study was to obtain further information about the causes of disturbed testicular descent on the one hand, and on the other to gain fresh knowledge about the regulatory mechanisms of testicular descent from such a hormone-induced disturbance of descent. The assumption that a change in the proportions of the genitals can be related to disturbances of descent could not be confirmed.

The qualitative and quantitative morphological evaluation revealed that distinct inhibition of testicular descent associated with suppression of the gubernacula occurs in the male fetuses of maternal animals treated with ethinyl estradiol, and to an even greater extent in those of maternal animals treated with a combination of ethinyl estradiol and cyproterone acetate. However, treatment with cyproterone acetate alone was also not without its consequences – at least, the testes were not in a position in keeping with the age of the animals, close up against the neck of the bladder. Differentiation of the ligamentous apparatus was, however, normal. The further, postnatal course of descent was tardy in all three treated groups and terminated in a normal descent only in group II (cyproterone acetate), the other two groups displaying disturbances of descent.

A comparison of the morphological data at the time of birth with those from the first few postnatal weeks gave rise to the hypothesis that the regulatory mechanism of this process cannot be elucidated simply on the basis of androgen dependence or Factor X dependence — rather it is suggested that Factor X is primarily responsible for differentiation of the ligamentous of the gonads, while androgens either alone or in addition are required for the final phase of descent.

*Acknowledgement.* The authors wish to thank Prof. G. Aumüller, Institute of Anatomy and Cell Biology of the Philipps-University, Marburg, for his valuable contribution.

# References

Bergh A, Damber JE (1979) Morphological and endocrinological differences between the abdominal testes in bilateral and unilateral cryptorchid rats. Int J Androl 2:319–329

Bergin WC, Gier HT, Marion GB, Coffman JR (1970) A developmental concept of equine cryptorchidism. Biol Reprod 3:82–92

Bierich JR (1977) Treatment by human chorionic gonadotrophin in maldescended testes. In: Bierich JR, Rager K, Ranke MB (eds) Maldescensus testis. Urban and Schwarzenberg, München, pp 101–109

Blanchard MG, Josso N (1974) Source of the anti-Müllerian hormone synthesized by the fetal testis: müllerian-inhibiting activity of fetal bovine Sertoli cells in tissue culture. Pediat Res 8:968–971

Bouin P, Ancel P (1903) Sur la signification de la glande interstitielle du testicule embryonnaire. C R Soc Biol (Paris) 55:1682–1684

Budzik GP, Swann DA, Hayashi A, Donahoe PK (1980) Enhanced purification of müllerian inhibiting substance by lectin affinity chromatography. Cell 21:909–915

Damber JE, Bergh A, Janson PO (1978) Testicular blood flow and testosterone concentration in the spermatic venous blood in rats with experimental cryptorchidism. Acta Endocrinol (Copenh) 88:611–618

Döhler KD, Wuttke W (1975) Changes with age in levels of serum gonadotrophin, prolactin, and gonadal steroids in prepubertal male and female rats. Endocrinology 97:898–907

Donahoe PK, Swann DA (1977) The role of müllerian inhibiting substance in mammalian sex differentiation. In: Johnson MH (ed) Development in mammals, vol 2. North-Holland, New York, pp 323–335

Donahoe PK, Ito Y, Marfatia S, Hendren W (1976) The production of the müllerian inhibiting substance by the fetal, neonatal and adult rat. Biol Reprod 15:329–334

Elger W, Neumann F, Von Berswordt-Wallrabe R (1971) The influence of androgen antagonists and progestagens on the sex differentiation of different mammalian species. In: Hamburgh M, Barrington EJW (eds) Hormones in development, chap 51. Appleton-Century-Crofts. Educational Division, Meredith Corporation, New York, pp 641–667

Elger W, Richter J, Korte R (1977) Failure to detect androgen dependence of the descensus testiculorum in foetal rabbits, mice and monkeys. In: Bierich JR, Rager K, Ranke MB (eds) Maldescensus testis. Urban and Schwarzenberg, Baltimore, pp 187–190

Forest MG (1977) Plasma steroid hormones in prepubertal cryptorchid boys in basal state and after long-term gonadotropin stimulation. In: Bierich JR, Rager K, Ranke MB (eds) Maldescensus testis. Urban and Schwarzenberg, Baltimore, pp 69–78

Frowein J, Engel W (1975) Binding of human chorionic gonadotrophin by rat testis: effect of sexual maturation, cryptorchidism and hypophysectomy. J Endocrinol 64:59–66

Gier HT, Marion GB (1969) Development of mammalian testes and genital ducts. Biol Reprod 1:1–23

Gier HT, Marion GB (1970) Development of the mammalian testis. In: Johnson AD, Gomes WR (eds) The testis, vol 1. Academic, New York, p 1

Greene RR, Burrill MW, Ivy AC (1939) Experimental intersexuality. The effect of antenatal androgens on sexual development of female rats. Am J Anat 65:415–469

Gupta D (1979) Endocrinology data in experimental cryptorchidism in the rat. Pediat Adolesc Endocrinol 6:64–78

Gupta D, Rager K (1977) Biogenesis of androgens in normal and cryptorchid testes. In: Bierich JR, Rager K, Ranke MB (eds) Maldescensus testis. Urban and Schwarzenberg, München, pp 57–67

Gupta D, Zarzycki J, Rager K (1975) Plasma testosterone and dihydrotestosterone in male rats during sexual maturation and following orchidectomy and experimental bilateral cryptorchidism. Steroids 25:33–42

Hadziselimović F (1977) Cryptorchidism. Adv Anat Embryol Cell Biol 53:5–77

Hadziselimovic F. Herzog B (1976) The meaning of the Leydig cell in relation to the etiology of cryptorchidism. J Pediatr Surg 11:1–8

Hadziselimovic F, Herzog B (1977) Development of normal and cryptorchid human testes an ultrastructural study. In: Bierich JR, Rager K, Ranke MB (eds) Maldescensus testis. Urban and Schwarzenberg, Baltimore, pp 39–46

Hadziselimovic F, Girad J, Herzog B (1977) Treatment of cryptorchidism by synthetic luteinising-hormone-releasing hormone. Lancet 26:1125

Hall RW, Gomes WR (1975) The effect of artificial cryptorchidism on serum oestrogen and testosterone levels in the adult male rat. Acta Endocrinol (Copenh) 80:583–591

Hansson V, Weddington SC, Naess O, Attramadal A, French FS, Kotite N, Nayfeh SN, Ritzen EM, Hagenas L (1975) Testicular androgen binding protein (ABP) – a parameter of Sertoli cell secretory function. In: French FS, Hansson V, Ritzen EM, Nayfeh SN (eds) Hormonal regulation of spermatogenesis. Plenum, New York, p 323

Happ J, Neubauer M, Kollmann F, Egri A, Krawehl C, Demisch N, Beyer J (1975) Gonadotropin releasing hormone therapy in males with hypogonadotropic hypogonadism and in boys with maldescended testes. Acta Endocrinol (Copenh) [Suppl] 199:266 (Abstract)

Happ J, Kollmann F, Krawehl C, Depper M, Neubauer M, Krause U, Demisch K, Beyer J (1977) Gonadotropin and testosterone secretion in cryptorchid boys under long-term treatment with gonadotropin releasing hormone (GnRH). In: Bierich JR, Rager K, Ranke MB (eds) Maldescensus testis. Urban and Schwarzenberg, Baltimore, pp 133–136

Hedinger C (1979) Histological data in cryptorchidism. In: Cryptorchidism, diagnosis and treatment. Pediat Adolesc Endocrinol 6:3–13

Hillard MA, Bindon BM (1979) Plasma LH patterns in cryptorchid rams and wethers. J Reprod Fertil 43:379–380

Jean C, André M, Jean C, Berger M, De Turckheim M, Veyssiére G (1975) Estimation of testosterone and androstenedione in the plasma and testes of cryptorchid offspring of mice treated with oestradiol during pregnancy. J Reprod Fertil 4:235–247

Jones TM, Anderson W, Fang VS, Landau RL, Rosenfield RL (1977) Experimental cryptorchidism in adult male rats: histological and hormonal sequelae. Anat Rec 189:1–28

Josso N (1971) Interspecific character of the müllerian-inhibiting substance: action of the human fetal testis, ovary and adrenal on the fetal rat müllerian duct in organ culture. J Clin Endocrinol Metab 32:404–409

Josso N (1972) Activité inhibitrice du testicule de foetus de vean sur le canal müller de foetus de rat, in culture organotypique: role des tuves seminiferes. C R Seances Acad Sci [III] 274: 3573–3576

Josso N (1973) In vitro synthesis of müllerian-inhibiting hormone by seminiferous tubules isolated from the calf fetal testis. Endocrinology 93:829–834

Josso N (1974a) Müllerian-inhibiting activity of human fetal testicular tissue deprived of germ cells by in vitro irradiations. Pediatr Res 8:755–758

Josso N (1974b) Fetal sexual differentiation. Pediatr Ann 3:67–79

Josso N (1977) Development and descent of the fetal testis. In: Bierich JR, Rager K, Ranke MB (eds) Maldescensus testis. Urban and Schwarzenberg, Baltimore, pp 3–12

Josso N, Briard ML (1980) Embryonic testicular regression syndrome: variable phenotypic expression in siblings. J Pediat 97:200–204

Josso N, Tran D (1979) Biochemical aspects of prenatal testicular development: relationship to testicular descent. Pediat Adolesc Endocrinol 6:37–46

Josso N, Picard JY, Tran D (1977) The antimüllerian hormone. In: Recent Prog Horm Res 33: 117–157

Jost A (1947) Recherches sur la différenciation sexuelle de l'embryon de lapin. Rôle des gonades foetale dans la différénciation sexuelle somatique. Arch Anat Microsc Morphol Exp 36:151–315

Jost An (1948) Activité androgène du testicule de rat greffé sur l'adulte castré. C R Soc Biol (Paris) 142:196–198

Jost A (1967) Steroids and sex differentiation of the mammalian foetus. In: Martini L, Fraschini F, Motta M (eds) Proceedings of the second international congress on hormonal steroids. Excerpta Medica Foundation, Amsterdam pp 74–81

Jost A (1970) Hormonal factors in the sex differentiation of the mammalian foetus. Philos Trans R Soc Lond [Biol] 259:119–130

Keel BA, Abney TO (1980) Influence of bilateral cryptorchidism in the mature rat: alterations in testicular function and serum hormone levels. Endocrinology 107:1226–1233

Keller K (1922) Über Geschlechtstransformation beim Säugetier. Betrachtungen über die Entstehung der Geschlechtsmißbildung beim unfruchtbaren Zwilling des Rindes. Wien Tierärztl Monatsschr 9:193–204

Keller K, Tandler J (1916) Über das Verhalten der Eihäute bei der Zwillingsträchtigkeit des Rindes. Untersuchungen über die Entstehungsursache der geschlechtlichen Unterentwicklung von weiblichen Zwillingskälbern, welche neben einem männlichen Kalb zur Entwicklung gelangen. Wien Tierärztl Monatsschr 3:513–526

Knobil E (1980) The neural control of the menstrual cycle. Recent Progr Horm Res 36:53–88

Knorr D, Pröschold, Richter W (1977) Fertility after treatment of maldescensus testis. In: Bierich JR, Rager K, Ranke MB (eds) Maldescensus testis. Urban and Schwarzenberg, München, pp 95–99

Kollmann F, Happ J, Krawehl C, Neubeuer M, Usadel KA, Leitner O, Sandow J (1977) Treatment of cryptorchidism with LH-RH nasal spray. In: Bierich JR, Rager K, Ranke MB (eds) Maldescensus testis. Urban and Schwarzenberg, Baltimore, pp 141–144

Lee PA, Hoffman WH, White WH, Engel RME, Blizzard RM (1974) Serum gonadotropins in cryptorchidism. Am J Dis Child 127:530–532

Lillie FR (1916) The theory of the freemartin. Science 43:611–613

Lillie FR (1917) The freemartin, a study of action of the sex hormones in the foetal life of cattle. J Exp Zool 23:371–451

Lipshultz LI (1976) Cryptorchidism in the subfertile male. Fertil Steril 27:609–620

McLachlan JA, Newbold RR (1975) Reproductive tract lesions in male mice exposed prenatally to diethylstilbestrol. Science 190:991–992

Miyachi Y, Nieschlag E, Lipsett MB (1973) The secretion of gonadotropins and testosterone by the neonatal rat. Endocrinology 92:1–5

Moore CR (1951) Experimental studies on the male reproductive system. J Urol 65:497–499

Moore CR, Oslund R (1924) Experiments on the sheep testis cryptorchidism, vasectomy and scrotal insulation. Am J Physiol 67:595–597

Mühlenstedt D, Schneider HPG (1979) Sexual differentiation of the hypothalamus in gonadal agenesis and testicular feminization. Arch Gynecol 227:97–102

Nelson WO (1951) Mammalian spermatogenesis: effect of experimental cryptorchidism in the rat and non-descent of the testis in man. Recent Progr Horm Res 6:29–62

Neumann F, Kramer M (1964) Antagonism of androgenic and antiandrogenic agents in their action on the rat fetus. Endocrinology 75:428–433

Neumann F, Hamada H (1963) Intrauterine Feminisierung männlicher Rattenfoeten durch das stark gestagen wirksame 6-chlor-1,2-methylen-17-hydroxy-progesteron-acetat. In: Schilddrüsenhormone und Körperperipherie-Regulation der Schilddrüsenfunktion. Springer, Berlin Göttingen Heidelberg, pp 301–304

Neumann F, Richter KD, Günzel P (1965) Wirkungen von Antiandrogenen. Zentralbl Veterinärmed [A] 12:171–188

Neumann F, Hahn JD, Kramer M (1967) Hemmung von testosteronabhängigen Differenzierungsvorgängen der männlichen Ratte nach der Geburt. Acta Endocrinol 54:227–240

Neumann F, Elger W, Steinbeck H (1969) Drug induced intersexuality in mammals. J Reprod Fertil [Suppl] 7:9–24

Neumann F, Steinbeck H, Elger W (1970) Sexualdifferenzierung. (Sexual differentiation, morphology of development and maturation of endocrine organs). In: Kracht J (ed) Endokrinologie der Entwicklung und Reifung. Springer, Berlin Heidelberg New York, pp 58–82

Nomura T, Masuda M (1980) Carcinogenic and teratogenic activities of diethylstilbestrol in mice. Life Sci 26:1955–1962

Odell WD, Swerdloff RS (1976) Male hypogonadism. Med Prog Technol 124:446–475

Picard JY, Tran D, Josso N (1978) Biosynthesis of labelled anti-Müllerian hormone by fetal testes: evidence for the glycoprotein nature of the hormone and for its disulfide-bonded structure. Mol Cell Endocrinol 12:17–30

Portman A (ed) (1948) Einführung in die vergleichende Morphologie der Wirbeltiere. Benno Schwabe, Basel, pp 13–329

Prader A, Illig R, Zachmann M (1976) Prenatal LH-deficiency as a possible cause of male pseudohermaphroditism, hypospadias, hypogenitalism and cryptorchidism. Pediatr Res 10:883 (Abstract)

Prasad MRN (1974) Männliche Geschlechtsorgane. In: Helmcke JG, Starck D, Wermuth H (eds) Handbuch der Zoologie, vol 8. Walter De Gruyter, New York, pp 19–35

Raynaud A (1940) Effets, sur la differénciation sexuelle des embryons, d'un mélange de dipropionate d'oestradiol et testosterone injecté à la souris en gestation. C R Acad Sci (Paris) 211:572–574

Raynaud A (1942) Modification expérimentale de la differénciation sexuelle des embryons de souris par action hormones androgènes et oestrogènes (Etude des d'intersexualite que en resultent). Actual Scient Ind (Paris) 925:1–202

Raynaud A (1957) Recherches sur les facteurs de la differénciation sexuelle de l'appareil gubernaculaire du foetus de souris. C R Acad Sci (Paris) 245:2393–2396

Raynaud A (1958) Inhibition, sous l'effet d'une hormone oestrogene, du development du gubernaculum du foetus mâle de souris. C R Acad Sci (Paris) 246:176–178

Raynaud A, Frilley M (1947) Destruction des glandes génitales de l'embryon de souris par une irradiation au moyen des rayons X, l'âge de treize jours. Ann Endocrinol 8:141–164

Richter J (1973) Die hormonale Steuerung des Descensus testiculorum. Untersuchungen mit Androgenen, Antiandrogenen und Östrogenen an männlichen und weiblichen Mäusefötten. Dissertation, Freie Universität Berlin, pp 1–63

Ritzen EM, Hagenäs L, Purvis K, Gurrero R, Johnsonbaugh RE, Dym M, French FS, Hansson V (1977) Androgens and androgen binding protein (Abp) in testicular fluids. In: Bierich JR, Rager K, Ranke MB (eds) Maldescensus testis. Urban and Schwarzenberg, München, pp 79–87

Romer AS (1971) Exkretions- und Fortpflanzungsorgane. In: Vergleichende Anatomie der Wirbeltiere. Paul Parey, Hamburg, pp 330–369

Scorer CG (1964) The descent of the testis. Arch Dis Child 39:605–609

Scorer CG, Farrington GH (1971) Histological studies of the undescended testis. In: Congenital deformities of the testis and epididymis. Butterworth, London, pp 58–75

Starck D (1975) Entwicklung des Urogenitalsystems. In: Embryologie, part B, chap 3. Georg Thieme, Stuttgart, pp 500–532

Swerdloff RS, Patrick CW, Jacobs HS, Odell WD (1971) Serum LH and FSH during sexual maturation in the male rat: effect of castration and cryptorchidism. Endocrinology 88:120–127

Torrey TW (1945) The development of the urogenital system of the albino rat. Am J Anat 76: 27–31

Tran D, Meusy-Dessolle N, Josso N (1977) Anti-Müllerian hormone is a functional marker of foetal Sertolis cells. Nature 269:411–412

Waaler PE (1979) Morphometric studies in undescended testes. Pediat Adolesc Endocrinol 6:27–36

Wensing CJG (1968) Testicular descent in some domestic mammals. I. Anatomical aspects of testicular descent. Proc Kon Ned Akad Wetensch C 71:423–434

Wensing CJG (1973a) Testicular descent in some domestic mammals. III. Search for the factors that regulate the gubernacular reaction. Proc Kon Ned Akad Wetensch C 76:196–200

Wensing CJG (1973b) Abnormalities of testicular descent. Proc Kon Ned Akad Wetensch C 76: 373–381

Wensing CJG, Colenbrander B (1973) Cryptorchidism and inguinal hernia. Proc Kon Ned Akad Wetensch C 76:489–494

Wensing CJG, Colenbrander B (1977) The process of normal and abnormal testicular descent. In: Bierich JR, Rager K, Ranke MB (eds) Maldescensus testis. Urban and Schwarzenberg, Baltimore, pp 193–198

Wensing CJG, Colenbrander B, Bosma AAA (1975) Testicular feminization syndrome and gubernacular development in a pig. Proc Kon Ned Akad Wetensch C 78:402–405

Wilson JD (1975) Dihydrotestosterone formation in cultured human fibroblast. Comparison of cells from normal subjects and patients with familial incomplete male pseudohermaphroditism, type 2. Biol Chem 250:3498–3504

Wislocki GB (1933) Location of the testes and body temperature in mammals. Q Rev Biol (Baltimore) 8:385–396

Witschi E (1951) Gonadal development and function embryogenesis of the adrenal and the reproductive glands. Recent Progr Horm Res 6:1–27

Wolff E (1962) The effect of ovarian hormones on the development of the urogenital tract and mammary primordia. In: The ovary, vol II. Academic, New York, pp 155–178

# Subject Index

## The Pituitary and Testis

Clinical and Experimental Studies

Editors: D.M.de Kretser, H.G.Burger, B.Hudson
1983. 92 figures, 17 tables. XI, 186 pages
(Monographs on Endocrinology, Volume 25)
ISBN 3-540-11874-8

R. Volpé

## Auto-immunity
## in the Endocrine System

1981. 32 figures, 15 tables. X, 187 pages
(Monographs on Endocrinology, Volume 20)
ISBN 3-540-10677-4

## Vacricocele and Male Infertility

Recent Advances in Diagnosis and Therapy

Editors: E. Jecht, E. Zeitler
1982. 98 figures. XVI, 211 pages.
ISBN 3-540-10727-4

F. Hadžiselimović

## Cryptorchidism

Management and Implications

With contributions by W.J.Cromie, F.Hinman,
B.Höcht, S.J.Kogan, T.S.Trulock, J.R.Woodard
Foreword by F.Hinman
1983. 67 figures. XV, 135 pages
ISBN 3-540-11881-0

## Prostate Cancer

Editor: W.Duncan
1981. 68 figures, 67 tables. X, 190 pages
(Recent Results in Cancer Research, Volume 78)
ISBN 3-540-10676-6

## Gonadal Steroids and Brain Function

IUPS-Satellite-Symposium, Berlin, July 10–11, 1980

Editors: W. Wuttke, R. Horowski
1981. 136 figures, 10 tables. XIII, 373 pages
(Experimental Brain Research, Supplementum 3)
ISBN 3-540-10606-5

Springer-Verlag
Berlin
Heidelberg
New York
Tokyo

# Advances in Anatomy Embryology and Cell Biology

Editors: F. Beck, W. Hild,
J. van Limborgh, R. Ortmann,
J.E. Pauly, T.H. Schiebler

Springer-Verlag
Berlin
Heidelberg
New York
Tokyo

Volume 72: H. Breuker
**Seasonal Spermatogenesis in the Mute Swan (Cygnus olor)**
1982. 30 figures. VII, 94 pages. ISBN 3-540-11326-6

Volume 73: G. Zweers
**The Feeding System of the Pigeon (Columba livia L.)**
1982. 45 figures. VII, 108 pages. ISBN 3-540-11332-0

Volume 74: J. Altman, S.A. Bayer
**Development of the Cranial Nerve Ganglia and Related Nuclei in the Rat**
1982. 64 figures. VII, 90 pages. ISBN 3-540-11337-1

Volume 75: V. Grouls, B. Helpap
**The Development of the Red Pulp in the Spleen**
1982. 37 figures. 80 pages. ISBN 3-540-11408-4

Volume 76: P. Kugler
**On Angiotensin-Degrading Aminopeptidases in the Rat Kidney**
1982. 88 figures. 96 pages. ISBN 3-540-11452-1

Volume 77: E. Braak
**On the Structure of the Human Striate Area**
1982. 44 figures. XI, 87 pages. ISBN 3-540-11512-9

Volume 78: G. Grün
**The Development of the Vertebrate Retina: A Comparative Survey**
1982. 15 figures. VIII, 85 pages. ISBN 3-540-11770-9

Volume 79: S.F. Perry
**Reptilian Lungs**
Functional Anatomy and Evolution
1983. 32 figures. Approx. 80 pages. ISBN 3-540-12194-3

Volume 80: J. Koebke
**A Biomechanical and Morphological Analysis of Human Hand Joints**
1983. 50 figures. Approx. 100 pages. ISBN 3-540-12438-1